Revision Guide

Cambridge
International AS and A Level

Geography

Garrett Nagle & Paul Guinness

WITHDRAWN

HODDER
EDUCATION

To Angela, Rosie, Patrick and Bethany

Hodder Education, an Hachette UK company, 338 Euston Road, London NW1 3BH

Orders
Bookpoint Ltd, 130 Milton Park, Abingdon, Oxfordshire OX14 4SB
tel: 01235 827827
fax: 01235 400401
e-mail: education@bookpoint.co.uk
Lines are open 9.00 a.m.–5.00 p.m., Monday to Saturday, with a 24-hour message answering service. You can also order through the Hodder Education website: www.hoddereducation.co.uk

ISBN 978-1-4441-8148-7

First printed 2013
Impression number 5 4 3 2 1
Year 2018 2017 2016 2015 2014 2013

Cover photo by Chris Guinness

Typeset by Datapage (India) Pvt. Ltd.
Printed and bound by CPI Group (UK) Ltd, Croydon, CR0 4YY

This text has not been through the Cambridge endorsement process.

Hachette UK's policy is to use papers that are natural, renewable and recyclable products and made from wood grown in sustainable forests. The logging and manufacturing processes are expected to conform to the environmental regulations of the country of origin.

Get the most from this book

Everyone has to decide his or her own revision strategy, but it is essential to review your work, learn it and test your understanding. This Revision Guide will help you to do that in a planned way, topic by topic. Use this book as the cornerstone of your revision and don't hesitate to write in it — personalise your notes and check your progress by ticking off each section as you revise.

☑ **Tick to track your progress**

Use the revision planner on pages 4 and 5 to plan your revision, topic by topic. Tick each box when you have:

● revised and understood a topic
● tested yourself
● practised the exam-style questions

You can also keep track of your revision by ticking off each topic heading in the book. You may find it helpful to add your own notes as you work through each topic.

2.2 The global energy budget

2 Atmosphere and weather

Latitudinal variations in radiation Revised

Atmospheric energy

The atmosphere is an open energy system receiving energy from both Sun and Earth. Although the latter is very small, it has an important local effect, as in the case of urban climates. **In**coming **sol**ar radi**ation** is referred to as **insolation**.

There are important variations in the receipt of solar radiation with latitude and season (Figure 2.3). The result is an imbalance: positive budget in the tropics, negative one at the poles. However, neither region is getting progressively hotter or colder. To achieve this balance the horizontal transfer of energy from the equator to the poles takes place by winds and ocean currents. This gives rise to an important second energy budget in the atmosphere – the horizontal transfer

Features to help you succeed

Expert tip

Throughout the book there are tips from the experts on how to maximise your chances.

Definitions and key words

Clear, concise definitions of essential key terms are provided on the page where they appear.

Key words from the syllabus are highlighted in bold for you throughout the book.

Typical mistake

Advice is given on how to avoid the typical mistakes students often make.

Exam-style questions

Exam-style questions are provided for each topic. Use them to consolidate your revision and practise your exam skills.

Exam ready ☐

Now test yourself

These short, knowledge-based questions provide the first step in testing your learning. Answers are at the back of the book.

Tested ☐

My revision planner

Paper 1 Core Geography

Paper 2 Advanced Physical Geography Options

Countdown to my exams

6–8 weeks to go

- Start by looking at the syllabus — make sure you know exactly what material you need to revise and the style of the examination. Use the revision planner on pages 4 and 5 to familiarise yourself with the topics.

- Organise your notes, making sure you have covered everything on the syllabus. The revision planner will help you to group your notes into topics.

- Work out a realistic revision plan that will allow you time for relaxation. Set aside days and times for all the subjects that you need to study, and stick to your timetable.

- Set yourself sensible targets. Break your revision down into focused sessions of around 40 minutes, divided by breaks. This Revision Guide organises the basic facts into short, memorable sections to make revising easier.

Revised ☐

4–6 weeks to go

- Read through the relevant sections of this book and refer to the expert tips, typical mistakes and key terms. Tick off the topics as you feel confident about them. Highlight those topics you find difficult and look at them again in detail.

- Test your understanding of each topic by working through the 'Now test yourself' questions in the book. Look up the answers at the back of the book.

- Make a note of any problem areas as you revise, and ask your teacher to go over these in class.

- Look at past papers. They are one of the best ways to revise and practise your exam skills. Write or prepare planned answers to the exam-style questions provided in this book. Check your answers with your teacher.

- Try different revision methods. For example, you can make notes using mind maps, spider diagrams or flash cards.

- Track your progress using the revision planner and give yourself a reward when you have achieved your target.

Revised ☐

1 week to go

- Try to fit in at least one more timed practice of an entire past paper and seek feedback from your teacher, comparing your work closely with the mark scheme.

- Check the revision planner to make sure you haven't missed out any topics. Brush up on any areas of difficulty by talking them over with a friend or getting help from your teacher.

- Attend any revision classes put on by your teacher. Remember, he or she is an expert at preparing people for examinations.

Revised ☐

The day before the examination

- Flick through this Revision Guide for useful reminders, for example the expert tips, typical mistakes and key terms.

- Check the time and place of your examination.

- Make sure you have everything you need — extra pens and pencils, tissues, a watch, bottled water, sweets.

- Allow some time to relax and have an early night to ensure you are fresh and alert for the examinations.

Revised ☐

My exams

Paper 1

Date: ....

Time: ....

Location:...

Paper 2

Date: ....

Time: ....

Location:...

Paper 3

Date: ....

Time: ....

Location:...

1 Hydrology and fluvial geomorphology

1.1 The drainage basin system

The drainage basin system

Revised

A **drainage basin** is a natural system with inputs, flows and stores of water and sediment. Every drainage basin is unique, on account of its climate, geology, vegetation, soil types, size, shape and human activities. The drainage basin system is an **open system** as it allows the movement of energy and matter across its boundaries.

The **hydrological cycle** refers to the cycle of water between atmosphere, lithosphere and biosphere. Water can be stored at a number of stages or levels within the cycle (Figure 1.1). These stores include vegetation, surfaces, soil moisture, groundwater and water channels. Human modifications to these can be made at every scale.

> A **drainage basin** refers to the area drained by a river and its tributaries.
>
> **Hydrology** is the study of water as it moves on, and under and through the Earth's surface.
>
> The **water cycle** or **hydrological cycle** is the movement of water between air, land and sea.

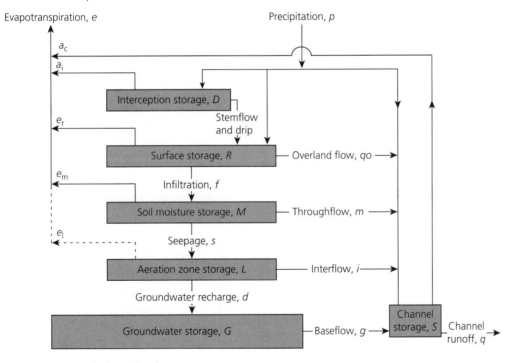

Figure 1.1 The hydrological cycle

- **Precipitation** is the main input into the drainage basin system. The main characteristics that affect local hydrology include: the total amount of precipitation; intensity (mm/hour); type of precipitation (snow, rain etc.); geographical distribution; and temporal variability (i.e. seasonality).
- **Interception** has three main components: **interception loss** – water that is retained by plant surfaces and which is later evaporated away or absorbed by the plant; **throughfall** – water that either falls through gaps in the vegetation or which drops from leaves, twigs or stems; **stemflow** – water that trickles along twigs and branches and finally down the main trunk.

> **Precipitation** is the conversion and transfer of moisture in the atmosphere to the land. It includes rainfall, snow, frost, hail and dew.
>
> **Interception** is the precipitation that is collected and stored by vegetation.

- **Evaporation** and **transpiration** increase under warm, dry conditions and decreases under cold, calm conditions. Meteorological factors affecting **evapotranspiration** (EVT) include temperature, humidity and windspeed. Of these, temperature is the most important. Other factors include the amount of water available, vegetation cover and colour of the surface (albedo or reflectivity of the surface).

- EVT represents the most important aspect of water loss, accounting for the loss of nearly 100% of the annual precipitation in arid areas and 75% in humid areas.

- Potential evapotranspiration (PEVT) is the water loss that would occur if there was an unlimited supply of water in the soil for use by the vegetation. For example, the actual evapotranspiration rate in Egypt is less than 250 mm, because there is less than 250 mm of rain annually. However, given the high temperatures experienced in Egypt, if the rainfall was as high as 2000 mm, there would be sufficient heat to evaporate that water. Hence the potential evapotranspiration rate there is 2000 mm.

- The **infiltration capacity** is the maximum rate at which rain can be absorbed by a soil in a given condition. Infiltration is inversely related to **runoff** and is influenced by a variety of factors such as duration of rainfall, antecedent soil moisture (pre-existing levels of soil moisture), soil porosity, vegetation cover, raindrop size and slope angle.

Table 1.1 Influence of ground cover on infiltration rate

Ground cover	Infiltration (mm/hour)
Old permanent pasture	57
Permanent pasture: moderately grazed	19
Permanent pasture: heavily grazed	13
Strip-cropped	10
Weeds or grain	9
Clean tilled	7
Bare, crusted ground	6

- **Soil moisture** refers to the subsurface water in the soil. **Field capacity** refers to the amount of water held in the soil after excess water drains away, i.e. saturation or near saturation. **Wilting points** refer to the range of moisture content in which permanent wilting of different plants occurs. They define the approximate limits to plant growth. **Throughflow** refers to water flowing through the soil in natural pipes and **percolines** (lines of concentrated water flow between soil horizons).

- **Groundwater** refers to subsurface water. Groundwater accounts for 96.5% of all freshwater on the Earth. The permanently saturated zone within solid rocks and sediments is known as the phreatic zone. The upper layer of this is known as the **water table**. **Baseflow** refers to the part of a river's discharge that is provided by groundwater seeping into the bed of a river. It is a relatively constant flow, although it increases slightly following a wet period.

- **Recharge** refers to the refilling of water in pores where the water has dried up or been extracted by human activity. Hence, in some places, where **recharge** is not taking place, groundwater is considered a non-renewable resource.

- **Aquifers** are rocks that contain significant quantities of water. A **spring** is a natural flow of water from the Earth's surface. It occurs when the water table (the upper surface of saturation within permeable rocks) occurs at the surface.

Evaporation refers to the transformation of liquid water from the Earth's surface into a gas (water vapour).

Transpiration is water loss from vegetation to the atmosphere.

Evapotranspiration is the combined loss of water to the atmosphere through transpiration and evaporation.

Infiltration is the process by which water soaks into, or is absorbed by, the soil.

Runoff is water that flows over the land's surface.

Typical mistake

Drainage basin hydrology is very variable from year to year. This can be due to natural changes or, increasingly, human-related activities.

Expert tip

You may be asked to draw a diagram of a drainage basin hydrological cycle. A systems diagram – with inputs, stores, flows and outputs – such as that in Figure 1.1, is much better than one that tries to show trees, clouds, rainfall, glaciers, rivers, lakes and oceans, for example.

Now test yourself

1. Define the following hydrological characteristics:
   (a) interception
   (b) evaporation
   (c) infiltration
2. Study Figure 1.1.
   Outline the differences between overland flow, throughflow and baseflow.
3. Suggest what is meant by interception storage in Figure 1.1.
4. Outline what may happen to water that is stored on the surface on the ground.
5. Comment on the influence of ground cover on infiltration rates (Table 1.1).

Answers on p.213

Tested

1.2 Rainfall–discharge relationships within drainage basins

Annual hydrograph

Revised

A hydrograph is a line graph showing how water level in a river changes over time. There are two main types of hydrograph – annual hydrographs (also known as **river regimes**) and storm hydrographs (also known as **flood hydrographs**). Annual hydrographs show variations in the flow of a river over the course of a year, whereas a storm hydrograph shows the variation in the flow of a river for a period of between 1 and 7 days.

Stream flow occurs as a result of runoff, groundwater springs and input from lakes and from meltwater in mountainous or sub-polar environments. The character or **regime** of the resulting stream or river is influenced by several variable factors:

- the amount and nature of precipitation
- the local rocks, especially porosity and permeability
- the shape or morphology of the drainage basin, its area and slope
- the amount and type of vegetation cover
- the amount and type of soil cover

In Figures 1.2–1.4 discharge is shown in litres per second per km². On an annual basis the most important factor determining stream regime is climate. Figure 1.2 shows a river regime for the Guadalquivir river at Alcala Del Ri in Spain. Its peak flow is about 20 litres/second/km² in March. It is generally in higher flow during winter whereas in summer it has low flow. In fact, in August there appears to be no flow. Discharge in July and September is less than about 5 litres/second/km². This is due to the high-pressure system that characterises Mediterranean regions in summer, producing a summer drought. In contrast, winters are associated with low-pressure systems and the resulting rain they bring, hence higher discharges.

> A **flood hydrograph** shows how the discharge of a river varies over a short time – normally it refers to an individual storm or group of storms of not more than a few days in duration.
>
> A **river regime** is the annual variation in the flow of a river.

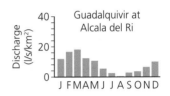

Figure 1.2 Guadalquivir regime

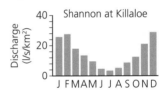

Figure 1.3 Shannon regime

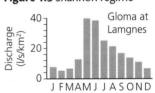

Figure 1.4 Gloma regime

Now test yourself

Tested

6 Study Figures 1.3 and 1.4. Compare the river regimes of the Gloma (Norway) and Shannon (Ireland).

7 Suggest reasons for their differences.

Answers on p.213

Expert tip

Make sure that you use units – it would be easy here to just refer to high discharge and low discharge (or high flow and low flow). A scale is provided – please make sure that you make use of it.

Flood hydrograph

Revised

A **flood hydrograph** (Figure 1.5) normally refers to an individual storm or group of storms of not more than a few days in length. Before the storm starts the main supply of water to the stream is through groundwater flow or **baseflow**. This is the main supplier of water to rivers. During the storm some water infiltrates into the soil while some flows over the surface as overland flow or runoff. This reaches the river quickly as **quickflow**. This causes the rapid rise in the level of the river. The **rising limb** shows us how quickly the floodwaters begin to rise, whereas the **recessional limb** is the speed with which the water level in the river declines after the peak. The **peak flow** is the maximum discharge of the river as a result of the storm and the **lag time** is the time between the height of the storm (not the start or the end) and the maximum flow in the river.

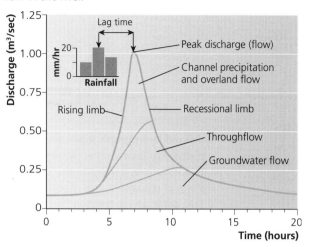

Figure 1.5 A simple hydrograph

Typical mistake

Some students work out the lag time from the start of the storm to the peak discharge. This is wrong – it should be from the peak of the storm to the peak of the flood.

The effect of urban development on hydrographs is to increase peak flow and decrease lag time (Figure 1.6). This is due to an increase in the proportion of impermeable ground in the drainage basin as well as an increase in the drainage density. Storm hydrographs also vary with a number of other factors (Table 1.2) such as basin shape, drainage density and gradient.

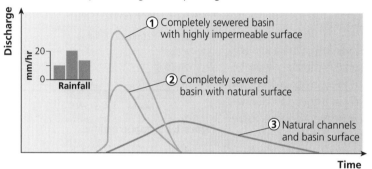

Source: *Advanced Geography: Concepts & Cases* by P. Guinness & G. Nagle (Hodder Education, 1999), p.255

Figure 1.6 The effects of urban development on flood hydrographs

Typical mistake

You may be expected to draw a labelled hydrograph – or hydrographs – to show the impact of urbanisation. Pay particular attention to the command words in the question – are you asked to describe the differences *or* are you being asked to explain the curves? Not reading the question (and therefore answering it) is one of the most common mistakes in exams.

Expert tip

Practise drawing annotated (labelled) diagrams of simple rural and urban hydrographs for the same storm. Label the rising and recessional limbs, peak flow and lag time.

Table 1.2 Factors affecting storm hydrographs

Factor	Influence on flood hydrograph
Precipitation type and intensity	Highly intensive rainfall is likely to produce overland flow and a steep rising limb and high peak flow. Low-intensity rainfall is likely to infiltrate into the soil and percolate slowly into the rock, thereby increasing the lag time and reducing the peak flow. Precipitation that falls as snow sits on the ground until it melts. Sudden, rapid melting can cause flooding and lead to high rates of overland flow, and high peak flows.
Temperature and evapotranspiration	Not only does temperature affect the type of precipitation, it also affects the evaporation rate (higher temperatures lead to more evaporation and so less water getting into rivers). On the other hand, warm air can hold more water, so the potential for high peak flows in hot areas is raised.
Antecedent moisture	If it has been raining previously and the ground is saturated or near saturated, rainfall will quickly produce overland flow and a high peak flow and short time lag.
Drainage basin size and shape	Smaller drainage basins respond more quickly to rainfall conditions. For example, the Boscastle (UK) floods of 2004 drained an area of less than 15 km². This meant that the peak of the flood occurred soon after the peak of the storm. In contrast, the Mississippi River is over 3700 km long – it takes much longer for the lower part of the river to respond to an event that occurs in the upper course of the river. Circular basins respond more quickly than linear basins, where the response is more drawn out.
Drainage density	Basins with a high drainage density – such as urban basins with a network of sewers and drains – respond very quickly. Networks with a low drainage density have a very long lag time.

Table 1.2 (continued)

Factor	Influence on flood hydrograph
Porosity and impermeability of rocks and soils	Impermeable surfaces cause more water to flow overland. This causes greater peak flows. Urban areas contain large areas of impermeable surfaces. In contrast, rocks such as chalk and gravel are permeable and allow water to infiltrate and percolate. This reduces the peak flow and increases the time lag. Sandy soils allow water to infiltrate, whereas clay is much more impermeable and causes water to pass overland.
Slopes	Steeper slopes create more overland flow, shorter lag times and higher peak flows.
Vegetation type	Broad-leafed vegetation intercepts more rainfall, especially in summer, and so reduces the amount of overland flow and peak flow, and increases lag time. In winter, deciduous trees lose their leaves and so intercept less.
Land use	Land uses that create impermeable surfaces or reduce vegetation cover reduce interception and increase overland flow. If more drainage channels are built (sewers, ditches, drains) the water is carried to rivers very quickly. This means that peak flows are increased and lag times reduced.

Now test yourself

Tested

8 Define the terms river regime and flood hydrograph.
9 Study Figure 1.6 which shows the impact of urbanisation on flood hydrographs. Describe the differences in the relationship between discharge and time.

Answers on p.213

1.3 River channel processes and landforms

River processes

Revised

Transport

The load is transported downstream in a number of ways:

- The smallest particles (silts and clays) are carried in suspension as the **suspended load**.
- Larger particles (sands, gravels, very small stones) are transported in a series of 'hops' as the **saltated load**.
- Pebbles are shunted along the bed as the **bed** or **traction load**.
- In areas of calcareous rock, material is carried in **solution** as the dissolved load.

The load of a river varies with discharge and velocity.

> **Capacity** of a stream refers to the largest amount of debris that a stream can carry.
>
> **Competence** refers to the diameter of the largest particle that can be carried.

Deposition and sedimentation: Hjulstrom curves

There are a number of causes of deposition, such as:

- a reduction in gradient, which decreases velocity and energy
- a decrease in the volume of water in the channel
- an increase in the friction between water and channel

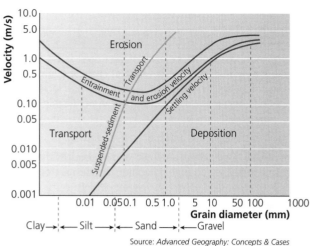

Source: *Advanced Geography: Concepts & Cases*
by P. Guinness & G. Nagle (Hodder Education, 1999), p.263

Figure 1.7 Hjulstrom curves

Hjulstrom curves show what work a river will do (erosion, transport, deposition) depending on its velocity and the size of material present. There are three important features of Hjulstrom curves:

- The smallest and largest particles require high velocities to lift them – for example, particles between 0.1 mm and 1 mm require velocities of around 100 mm/s to be entrained, compared with values of over 500 mm/s to lift clay (less than 0.01 mm) and gravel (over 2 mm). Clay resists entrainment due to its cohesion; gravel due to its weight.
- Higher velocities are required for entrainment than for transport.
- When velocity falls below a certain level those particles with a particular settling or fall velocity are deposited.

Erosion

- **Abrasion** (or **corrasion**) is the wearing away of the bed and bank by the load carried by a river. Abrasion increases as velocity increases.
- **Attrition** is the wearing away of the load carried by a river. It creates smaller, rounder particles.
- **Hydraulic action** is the force of air and water on the sides of rivers and in cracks.
- **Corrosion** or **solution** is the removal of chemical ions, especially calcium.

Velocity and discharge

- **Velocity** refers to the speed of the river. Velocities increase in rivers that are deeper and when rivers are in flood.
- **Discharge** refers to rate of flow of a river at a particular time. It is generally found by multiplying cross-sectional area by velocity, and is usually expressed in cusecs (cubic feet per second) or cumecs (cubic metres per second).
- The **hydraulic radius** is a measure of a stream's efficiency – it is calculated by dividing the cross-sectional area by the wetted perimeter (the length of bed and bank in contact with water). The higher the hydraulic radius, the more efficient the river.

Patterns of flow

There are three main types of flow: laminar, turbulent and helicoidal.

- For **laminar flow** a smooth, straight channel with a low velocity is required. This allows water to flow in sheets or laminae parallel to the channel bed.
- **Turbulent flow** occurs where there are higher velocities and an increase in bed roughness. Turbulence is associated with hydraulic action (cavitation). Vertical turbulence creates hollows in the channel bed.
- Horizontal turbulence often takes the form of **helicoidal flow** – a 'corkscrewing' motion. This is associated with the presence of alternating

pools and riffles in the channel bed, and where the river is carrying large amounts of material. The erosion and deposition by helicoidal flow creates meanders.

Channel landforms Revised ☐

Sinuosity is the length of a stream channel expressed as a ratio of the valley length. A low sinuosity has a value of 1.0 (i.e. it is straight), whereas a high sinuosity is above 4.4. The main groupings are **straight channels** (<1.5) and **meandering** (>1.5). Straight channels are rare. Even when they do occur the thalweg (line of maximum velocity) moves from side to side. These channels generally have a central ridge of deposited material, due to the water flow pattern. Meandering is a natural process and creates rivers with an asymmetric cross-section (Figure 1.8).

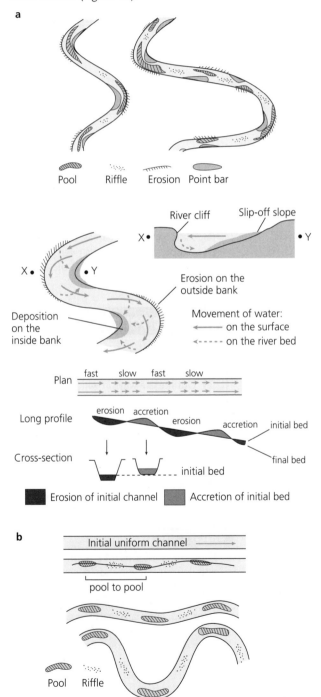

Figure 1.8 Characteristics of meanders

Braiding occurs when the channel is divided by islands or bars. Islands are vegetated and long-lived, whereas bars are unvegetated, less stable and often short-term features. Braided channels are formed by many factors, including:

- a large proportion of coarse material
- easily erodable bank material
- highly variable discharge

Pools are the deep parts of a meander, whereas **riffles** are the ridges of sediment found in the straighter sections of a meander. Riffles are generally formed of coarse gravel, whereas in a pool, erosion is the main process. Helicoidal flow in the river causes erosion on the outer bank of a meander, and the formation of **river cliffs**. In contrast, on the inner bank, helicoidal flow leads to the formation of **point bars** (also known as slip-off slopes). A **bluff** is another term for a river cliff, but may also be used to describe the edge of a river terrace (a former floodplain).

A **floodplain** is the part of a river valley, adjacent to the river channel, over which the river flows in times of flood. It is a zone of low relief and gentle gradients. The floodplain may contain oxbow lakes and is generally formed of alluvium.

Waterfalls occur where the river spills over a sudden change in gradient, undercutting rocks by hydraulic impact and abrasion, thereby creating a waterfall. The change in gradient is usually related to a band of resistant rock. A **gorge** is a deep, narrow valley with very steep sides. It is currently occupied by a river or had a river in the past.

Gorges may be formed as a result of:

- retreat of waterfalls (e.g. Niagara Falls)
- glacial overflow channelling
- collapse of underground caverns in carboniferous limestone areas
- surface runoff over limestone during a periglacial period

Rapids occur when there is a series of small bands of hard rock outcropping in a river. This causes turbulence within the river and a series of 'mini-waterfalls'. The Nile Cataracts are a good example of rapids.

Levées are formed when a river bursts its banks repeatedly over a long period of time. The floodwater quickly loses velocity, leading to the rapid deposition of coarse material (heavy and difficult to move a great distance) near the channel edge. These coarse deposits build up to form embankments, or **levées**. The finer material is carried further away to be deposited on the **floodplain**.

Alluvial fans are found in semi-arid areas where swiftly flowing mountain streams enter a main valley or plain at the foot of the mountains. There is a sudden decrease in velocity, causing deposition. Fine material is spread out as an alluvial fan, with a slope angle of less than 1°. By contrast, coarse material forms a relatively small, steep-sided **alluvial cone**, with a slope angle of up to 15°.

Deltas are river sediments deposited when a river enters a standing body of water such as a lake, a lagoon, a sea or an ocean. For a delta to form there must be a heavily laden river and a standing body of water with negligible currents, such as the Mediterranean or the Gulf of Mexico. Deposition is enhanced if the water is saline, because salty water causes small clay particles to flocculate or adhere together.

Typical mistake

Many students think that most erosion occurs in upland areas and deposition in lowland areas. Both occur throughout the course of a river. Most erosion only takes place when the river is in flood.

Now test yourself

10 Define the terms hydraulic action, attrition and abrasion.

11 Outline the ways in which a river transports its load.

12 State the approximate values needed to (a) deposit, (b) transport, (c) entrain and (d) erode a particle of 1.0 mm.

Answers on p.213

Tested ☐

Expert tip

For most landforms (of erosion and deposition) you should learn an annotated diagram, and ensure that you explain how the landform is formed. For many features, e.g. waterfalls and levées, you may need to learn a sequence of diagrams.

1.4 The human impact

Human impacts on hydrology — Revised

Urbanisation is a major cause of hydrological changes. These are summarised in Table 1.3.

Table 1.3 Potential hydrological effects of urbanisation

Urbanising influence	Potential hydrological response
Removal of trees and vegetation	Decreased evapotranspiration and interception; increased stream sedimentation
Initial construction of houses, streets and culverts	Decreased infiltration and lowered groundwater table; increased storm flows and decreased base flows during dry periods
Complete development of residential, commercial and industrial areas	Decreased porosity, reducing time of runoff concentration, thereby increasing peak discharges and compressing the time distribution of the flow; greatly increased volume of runoff and flood damage potential
Construction of storm drains and channel improvements	Local relief from flooding; concentration of floodwaters may aggravate flood problems downstream

Human impact on precipitation

There are a number of ways in which human activity affects precipitation. Cloud seeding has probably been one of the more successful. Rain requires particles, such as dust and ice, on which to form. Seeding introduces silver iodide, solid CO_2 (dry ice) or ammonium nitrate to attract water droplets.

Human impact on evaporation and transpiration

The human impact on evaporation and transpiration is relatively small in relation to the rest of the hydrological cycle but is nevertheless important.

Dams – there has been an increase in evaporation due to the construction of large dams. For example, Lake Nasser behind the Aswan Dam loses up to a third of its water due to evaporation. Water loss can be reduced by using chemical sprays on the surface, by building sand-fill dams and by covering the dams with plastic.

Urbanisation leads to a huge reduction in evapotranspiration due to the lack of vegetation. There may also be a slight increase in evaporation because of higher temperatures and increased surface storage.

Human impact on infiltration and soil water

Human activity has a great impact on infiltration and soil water. Land use changes are important. Urbanisation creates an impermeable surface, with compacted soil. This reduces infiltration and increases overland runoff and flood peaks. Infiltration is up to five times greater under forests compared with grassland.

Human impact on interception

Interception is determined by the density and type of vegetation. Most vegetation is not natural but represents some disturbance by human activity. Deforestation leads to:

- a reduction in evapotranspiration
- an increase in surface runoff
- a decline of surface storage
- a decline in lag time

Abstraction and water storage

In the High Plains of Texas, groundwater is now being used at a rapid rate to supply **centre-pivot irrigation schemes**. In under 50 years, the water level has declined by 30–50 m over a large area. The extent of the aquifer has reduced by more than 50% in large parts of certain counties. By contrast, in some industrial areas, recent reductions in industrial activity have led to less groundwater being taken out of the ground. As a result, groundwater levels in such areas have begun to rise, adding to the problem caused by leakage from ancient, deteriorating pipe and sewerage systems, and resulting in:

- surface water flooding
- pollution of surface waters and spread of underground pollution
- flooding of basements
- increased leakage into tunnels

There are various methods of **recharging** groundwater resources, providing that sufficient surface water is available. Where the materials containing the aquifer are permeable (as in some alluvial fans, coastal sand dunes or glacial deposits) water-spreading (through infiltration and seepage) is used.

> **Recharge** is the topping up of groundwater levels following abstraction.

Case study | **Impact of the Aswan Dam**

- Water losses – the dam provides less than half the amount of water expected due to increased evaporation from Lake Nasser.
- Salinisation – crop yields have been reduced by salinisation on up to one-third of the area irrigated by water from the dam.
- Groundwater changes – seepage leads to increased groundwater levels and may cause secondary salinisation.
- Displacement of population – up to 100,000 Nubian people were removed from their ancestral homes when the dam was constructed.
- Seismic stress – the earthquake of November 1981 is believed to have been caused by the dam. As water levels in the dam decrease so too does seismic activity.
- Deposition within the lake – infilling is taking place at a rate of about 100 million tonnes each year.

- Channel erosion (clear-water erosion) – because it has deposited its load behind the dam, water entering the river downstream of the dam has renewed ability to erode.
- Erosion of the Nile Delta – this is taking place at a rate of about 2.5 cm each year because of the reduced sediment load in the river.
- Loss of nutrients – the reduced sediment load means an estimated $100 million is needed to buy commercial fertilisers to make up for the lack of nutrients each year.
- Decreased fish catches – the loss of nutrients means that sardine yields are down 95% and 3000 jobs in Egyptian fisheries have been lost.
- Diseases have spread – such as schistosomiasis (bilharzia), which is more prevalent in the still, stagnant waters around the lake.

Drought

Revised

Drought is an extended period of dry weather leading to conditions of extreme dryness.

- Absolute drought is a period of at least 15 consecutive days with less than 0.2 mm of rainfall.
- Partial drought is a period of at least 29 consecutive days during which the average daily rainfall does not exceed 0.2 mm.

Dry conditions are caused by a number of factors.

- The main cause is the global atmospheric circulation. Dry, descending air associated with the **subtropical high-pressure belt** is the main cause of aridity around at 20–30° N and S.
- Distance from sea, or **continentality**, limits the amount of water carried across by winds.
- In other areas, **cold offshore currents** limit the amount of condensation in the overlying air.
- Some areas experience intense **rain-shadow effects**, as air passes over mountains.
- Human activities give rise to the spread of desert conditions into areas previously fit for agriculture. This is known as desertification, and is an increasing problem.

The effects of drought include hunger and malnutrition, declining crop yields, reduced water supplies for farming and other economic activities, stress on natural ecosystems, increased risk of fires and poor air quality.

Floods

Revised

Flooding occurs when a river overflows its banks. The main causes of floods are climatic forces, whereas conditions that intensify the flooding tend to be drainage basin specific. These flood intensifying conditions involve a range of human-related factors that alter the drainage basin response to a given storm:

- more rapid discharge in urban areas due to the impermeable surface and increased number of drainage channels
- urbanisation and urban growth (increase in impermeable surfaces)
- floodplain developments (increasing risk of damage)
- bridges, dams and obstructions (leads to ponding and possible flooding)
- changes in vegetation cover (e.g. agriculture)
- river engineering works (e.g. levées)
- human-induced climate change

Recurrence

The **recurrence interval** refers to the regularity of a flood of a given size. Small floods may be expected to occur regularly. Larger floods occur less often. A 100-year flood is a flood that is expected to occur, *on average*, once every 100 years. Increasingly larger floods are less common, but more damaging.

> **Typical mistake**
>
> Not all human impacts have the same effect on floods. Urban areas vary in terms of their degree of impermeable surfaces, the amount of open space and parkland etc.

> **Expert tip**
>
> Read the question carefully and pay attention to the command words – a question asking you to *describe* the causes of a flood expects a very different answer from a question that asks you to *explain* the causes of a flood.

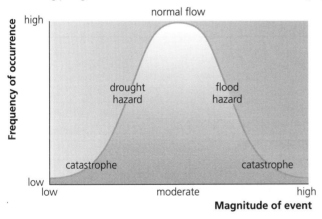

Figure 1.9 Floods and recurrence

Forecasting and warning

According to the United Nations Environment Programme's publication *Early Warning and Assessment* there are a number of things that could be done to improve flood warnings. These include:

- improved rainfall and snowpack estimates, and better and longer forecasts of rainfall
- better gauging of rivers, collection of meteorological information and mapping of channels
- better and current information about human populations, infrastructure, elevation and stream channels to improve flood risk assessment models
- better sharing of information between forecasters, national agencies, relief organisations and the general public
- more complete and timely sharing of meteorological and hydrological information among countries within international drainage basins
- sharing of technology among all agencies involved in flood forecasting and risk assessment, both within basins and throughout the world

Prevention and amelioration of floods

Loss sharing adjustments include disaster aid and insurance. **Disaster aid** refers to any aid, such as money, equipment, staff and technical assistance that is given to a community following a disaster. In developed countries **insurance** is an important loss-sharing strategy. However not all flood-prone households have insurance and many of those who are insured may be under insured.

Event modification adjustments include environmental control and hazard-resistant design. Physical control of floods depends on two measures – flood abatement and flood diversion. **Flood abatement** involves decreasing the amount of runoff, thereby reducing the flood peak in a drainage basin. There are a number of ways of reducing flood peaks, including:

- reforestation
- reseeding of sparsely vegetated areas to increase evaporative losses
- treatment of slopes, such as through contour ploughing or terracing, to reduce the runoff coefficient
- comprehensive protection of vegetation from wild fires, overgrazing and clear cutting
- clearance of sediment and other debris from headwater streams
- construction of small water and sediment holding areas
- preservation of natural water storage zones, such as lakes

Flood diversion measures, by contrast, include the construction of **levées**, **reservoirs**, and the modification of river channels. Levées are the most common form of river engineering. Reservoirs store excess rainwater in the upper drainage basin. However, this may only be appropriate in small drainage networks.

Now test yourself

13 What is meant by the term recurrence interval?

14 What is the difference between absolute and partial drought?

15 Where are the world's main dry areas to be found?

Answers on p.213

Tested

Exam-style questions

Section A

1 (a) Define the term hydrological cycle. [1]

(b) Draw an annotated diagram to show the main stores and flows in the hydrological cycle. [3]

2 Outline the ways in which human activities have modified the hydrological cycle. [6]

Section B

1 (a) Describe the ways in which a river carries its load. [3]

(b) Explain the main processes of erosion by a river. [4]

2 With the use of annotated diagrams, explain how waterfalls and ox-bow lakes are formed. [8]

3 With the use of examples, explain how human activities can increase the risk of flooding. [10]

Exam ready

2 Atmosphere and weather

2.1 Local energy budgets

The daytime energy budget
Revised

An **energy budget** refers to the amount of energy entering a system, the amount leaving the system, and the transfer of energy within the system. Energy budgets are commonly considered at a global scale (macro-scale) and at a local scale (micro-scale).

There are six components to the daytime energy budget – incoming solar radiation (insolation), reflected solar radiation, surface absorption, latent heat transfer (evaporation), sensible heat transfer and long-wave radiation. These influence the gain or loss of energy for a point at the Earth's surface. The daytime energy budget can be expressed by the formula:

energy available at the surface = incoming solar radiation − (reflected solar radiation + surface absorption + sensible heat transfer + long-wave radiation + latent heat transfers)

- **Incoming solar radiation (insolation)** is the main energy input and is affected by latitude, season and cloud cover. The less cloud cover there is, and/or the higher the cloud, the more radiation reaches the Earth's surface.
- **Reflected solar radiation (albedo)** varies with colour – light materials are more reflective than dark materials (Table 2.1). Grass has an average albedo of 20–30%, meaning that it reflects back about 20–30% of the radiation it receives.

> **Typical mistake**
>
> The term microclimate is sometimes used to describe regional climates, such as those associated with large urban areas, coastal areas or mountainous regions. Make sure you are clear about whether you are talking about a regional microclimate or a very small-scale microclimate.

> **Insolation** is the amount of incoming solar radiation (heat energy from the Sun) that reaches the Earth's surface.
>
> **Albedo** is the proportion of energy that is reflected back to the atmosphere.

Table 2.1 Selected albedo values

Surface	Albedo (%)
Water (Sun's angle over 40°)	2–4
Water (Sun's angle less than 40°)	6–80
Fresh snow	75–90
Old snow	40–70
Dry sand	35–45
Dark, wet soil	5–15
Dry concrete	17–27
Black road surface	5–10
Grass	20–30
Deciduous forest	10–20
Coniferous forest	5–15
Crops	15–25
Tundra	15–20

> **Now test yourself**
>
> Study Table 2.1.
> 1 What is meant by the term albedo?
> 2 Why is albedo important?
>
> **Answers on p.213**
>
> Tested

- **Surface absorption** occurs when energy reaches the Earth's surface, which heats up. How much it heats up depends on the nature of the surface. For example, if the surface can conduct heat to lower layers, the surface will remain cool. If the energy is concentrated at the surface, the surface warms up.

- **Sensible heat transfer** refers to the movement of parcels of air into and out from the area being studied. For example, air that is warmed by the surface may begin to rise (convection) and be replaced by cooler air. This is known as a convective transfer. It is very common in warm areas in the early afternoon.
- **Long-wave radiation** refers to the radiation of energy from the Earth (a cold body) into the atmosphere and, for some of it, eventually into space. There is, however, a downward movement of long-wave radiation from particles in the atmosphere. The difference between the two flows is known as the net radiation balance.
- **Latent heat transfer (evaporation)** occurs when heat energy is used to turn liquid water into water vapour. In contrast, when water vapour becomes a liquid, heat is released. Thus, when water is present at a surface, a proportion of the energy available will be used to evaporate it, and less energy will be available to raise local energy levels and temperature.

The night-time energy budget
Revised

The night-time energy budget consists of four components – long-wave radiation, latent heat transfer (condensation), absorbed energy returned to Earth (sub-surface supply), and sensible heat transfer.

- **Long wave radiation** – during a cloudless night, there is a large loss of long-wave radiation from the Earth. On a cloudy night, in contrast, the clouds return some long-wave radiation to the surface, hence the overall loss of energy is reduced.
- **Latent heat transfer (condensation)** is released when water condenses. During the night, water vapour in the air close to the surface can condense to form water, since the air has been cooled by the cold surface.
- **Sub-surface supply** refers to the heat transferred to the soil and bedrock during the day, which is released back to the surface at night. This can partly offset the night-time cooling at the surface.
- **Sensible heat transfer** refers to air movement. Cold air moving into an area may reduce temperatures whereas warm air may supply energy and raise temperatures.

> **Expert tip**
> Make a simple labelled diagram to show the daytime energy budget and the night-time energy budget.

Weather phenomena associated with local energy budgets
Revised

Mist and fog

Mist and fog are cloud at ground level.

- Mist occurs when visibility is between 1000 m and 5000 m and relative humidity is over 93%.
- Fog occurs when visibility is below 1000 m. Dense fog occurs when the visibility is below 200 m.

For mist and fog to form, condensation nuclei, such as dust and salt, are needed. These are more common in urban and coastal areas, so mist and fog are more common there.

For fog to occur, condensation must take place near ground level. Condensation can take place in two major ways:

- Air is cooled.
- More water is added to the atmosphere.

The cooling of air is quite common (orographic, frontal and convectional uplift). By contrast, the addition of moisture to the atmosphere is relatively rare. However, it does occur over warm surfaces such as the Great Lakes in North

> **Expert tip**
> Mist and fog form during calm, high-pressure conditions. If there is a low-pressure system, the winds will prevent fog from forming by mixing the air.

America or over the Arctic Ocean. Water evaporates from the relatively warm surface and condenses into the cold air above to form fog.

Contact cooling at a cold ground surface may produce saturation. As warm moist air passes over a cold surface it is chilled. Condensation takes place as the temperature of the air is reduced and the air reaches dew point (the temperature at which relative humidity is 100%). When warm air flows over a cold surface **advection fog** is formed. For example, near the Grand Banks off Newfoundland warm air from the Gulf Stream passes over the waters of the Labrador Current, which is 8–11°C cooler because it brings with it meltwater from the disintegrating pack-ice further north. This creates dense fog for 70–100 days each year.

Radiation fog occurs when the ground loses heat at night by long-wave radiation. This occurs during high-pressure conditions associated with clear skies.

Fog is a major environmental hazard – airports can be closed for many days and road transport is hazardous and slow. Freezing fog is particularly problematic. Large economic losses result from fog but the ability to do anything about it is limited.

Dew

Dew refers to condensation on a surface. The air is saturated, generally because the temperature of the surface has dropped enough to cause condensation. Occasionally, condensation occurs because more moisture is introduced, for example by a sea breeze, while the temperature remains constant.

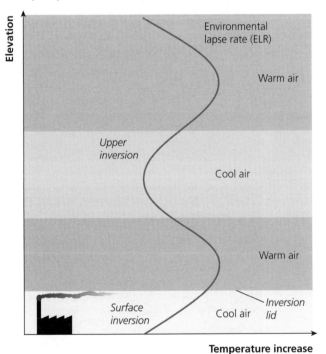

Figure 2.1 Temperature inversion

A relative increase in temperature with height in the lower part of the atmosphere is known as a **temperature inversion** (or radiation/nocturnal inversion – Figure 2.1). This happens when there are relatively calm (high-pressure) conditions and little mechanical turbulence from the wind causing the air to mix. As the cold air at the surface is dense, it will tend to stay at the surface. During the longer nights of winter there is even more time for the air near the surface to cool. During calm, high-pressure conditions the band of cooled air may extend a few metres before the warmer air is reached.

Temperature inversions are important as they influence air quality. Under high-pressure conditions and limited air movement, a temperature inversion will act like a lid on pollutants, causing them to remain in the lower atmosphere next to the Earth's surface (Figure 2.1). Only when the surface begins to heat up, and in turn warms the air above it, will the warm air be able to rise and with it any pollutants that it may contain.

Temperature inversions are common in depressions and valleys. Cold air may sink to the bottom of the valley and be replaced by warmer air above it. In some cases, the inversion can be so intense that frost hollows develop. These can reduce growth of vegetation, so are generally avoided by farmers. Urban areas surrounded by high ground, such as Mexico City and Los Angeles, are also vulnerable as cold air sinks from the mountains down to lower altitudes.

Now test yourself

Tested

5 Define the term 'temperature inversion'.

6 Explain why temperature inversions occur.

7 Describe the problems associated with temperature inversions.

Figure 2.2 shows rural and urban energy budgets for Washington DC (USA) during daytime and night-time. The figures represent the proportions of the original 100 units of incoming solar radiation dispersed in different directions.

a Rural surface

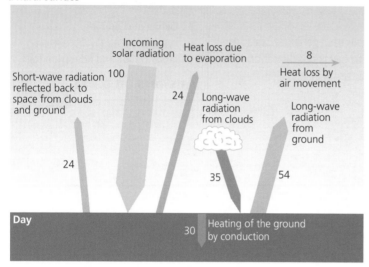
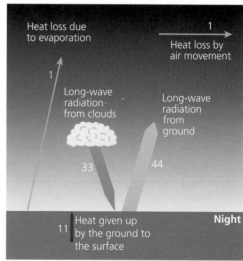

b Urban surface

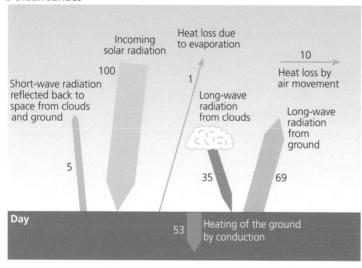
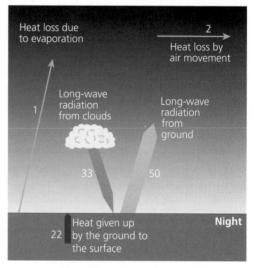

The figures represent the proportions of the original 100 units of incoming solar radiation dispersed in different directions.

Source: University of Oxford, 1989, Entrance examination for Geography

Figure 2.2 Daytime and night-time energy budgets for Washington DC

8 How does the amount of insolation received vary between the rural area and the urban area?

9 How does the amount of heat lost through evaporation vary between the areas?

10 Compare the amount of heat given up by the rural area and urban area at night. Suggest **two** reasons for these differences.

Answers on pp.213–214

2.2 The global energy budget

Latitudinal variations in radiation

Revised

Atmospheric energy

The atmosphere is an open energy system receiving energy from both Sun and Earth. Although the latter is very small, it has an important local effect, as in the case of urban climates. **In**coming **sol**ar radi**ation** is referred to as **insolation**.

There are important variations in the receipt of solar radiation with latitude and season (Figure 2.3). The result is an imbalance: positive budget in the tropics, negative one at the poles. However, neither region is getting progressively hotter or colder. To achieve this balance the horizontal transfer of energy from the equator to the poles takes place by winds and ocean currents. This gives rise to an important second energy budget in the atmosphere – the horizontal transfer between low latitudes and high latitudes to compensate for differences in global insolation.

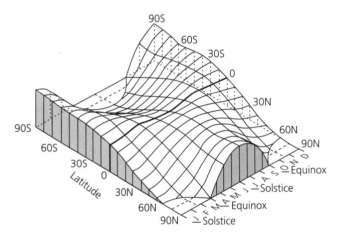

The variations of solar radiation with latitude and season for the whole globe, assuming no atmosphere. This assumption explains the abnormally high amounts of radiation received at the poles in summer, when daylight lasts for 24 hours each day.

Source: Barry, R. and Chorley, R., *Atmosphere, Weather and Climate*, Routledge, 1998

Figure 2.3 Seasonal and latitudinal variations in insolation

Expert tip

Make sure when talking about summer and winter in the southern or northern hemisphere that you refer to months – it is easy to forget that if it is summer in one hemisphere it is winter in the other.

Now test yourself

Tested

11 What does insolation stand for?

12 When does the South Pole receive most insolation (Figure 2.3)?

13 How much insolation does 80°N receive in December and January.

Answers on p.214

Pressure variations

Sea-level pressure conditions show marked differences between the hemispheres. In the northern hemisphere there are greater seasonal contrasts whereas in the southern hemisphere more stable average conditions exist (Figure 2.4). The differences are largely related to unequal distribution of land and sea, because ocean areas are much more equable in terms of temperature and pressure variations.

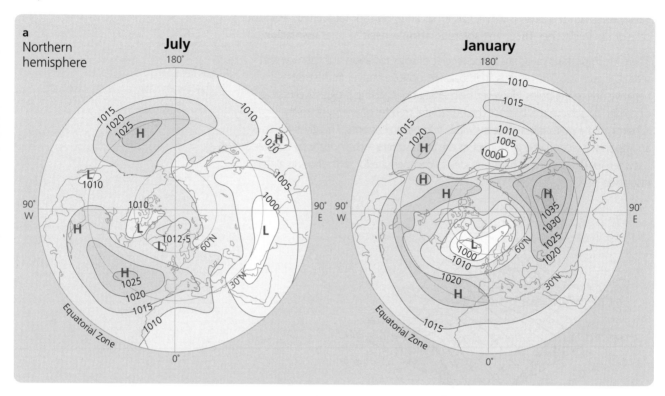

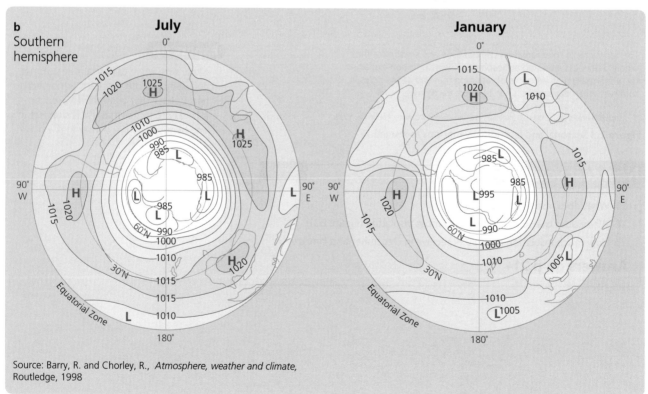

Source: Barry, R. and Chorley, R., *Atmosphere, weather and climate*, Routledge, 1998

Figure 2.4 Seasonal variations in pressure

Subtropical high-pressure belts (STHP) are a permanent feature, especially over ocean areas.

- In the southern hemisphere this is almost continuous at about 30° latitude.
- In the northern hemisphere, by contrast, at 30° the belt is much more discontinuous because of the land.
- Over the oceans high pressure occurs as discrete cells, such as the Azores and Pacific Highs.
- Over continental areas, such as southwest USA, southern Asia and the Sahara, major fluctuations occur: high pressure in winter and summer lows because of overheating.

Over the equatorial trough pressure is low, at around 1008–1010 mb.

- The trough coincides with the zone of maximum insolation.
- In the northern hemisphere in July it is well north of the equator (25° over India), whereas in the southern hemisphere (January) it is just south of the equator because land masses in the southern hemisphere are not of sufficient size to displace it southwards.

In temperate latitudes pressure is generally lower than in subtropical areas.

- The most unique feature is the large number of depressions (low pressure) and anticyclones (high pressure), which do not show up on a map of mean pressure.
- In the northern hemisphere there are strong winter low pressure zones over Icelandic and oceanic areas, but over Canada and Siberia high pressure dominates, due to the coldness of the land.
- In summer, high pressure is reduced, especially over continental areas.
- In polar areas pressure is relatively high throughout the year, especially over Antarctica, because of the coldness of the land mass.

Expert tip

Abbreviations are fine – for example LP for low pressure, SH for southern hemisphere – but when you first mention the term, write it out in full and add the abbreviation in brackets.

Now test yourself — Tested

14 What does STHP stand for?
15 Which has the greater seasonal contrast in pressure – land or sea?
16 How does a cold land mass influence pressure?
17 How does a warm land mass influence air pressure?

Answers on p.214

Surface wind belts — Revised

Winds between the tropics converge on a line known as the **inter-tropical convergence zone** (ITCZ) or equatorial trough (Figure 2.5).

The **inter-tropical convergence zone** is a band a few hundred kilometres wide in which winds from the tropics blow inwards, converge and then rise, forming an area of low-pressure.

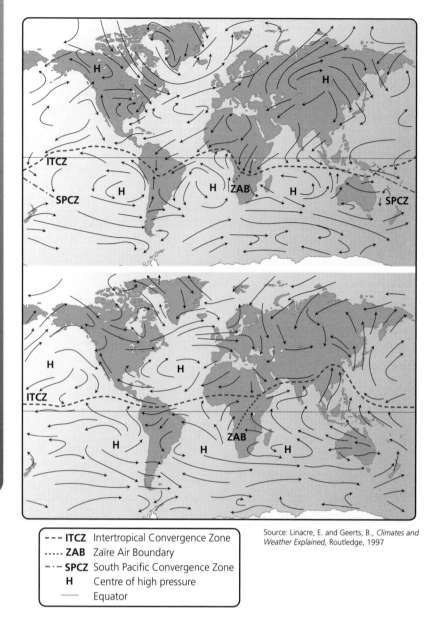

Source: Linacre, E. and Geerts, B., *Climates and Weather Explained,* Routledge, 1997

--- **ITCZ** Intertropical Convergence Zone
..... **ZAB** Zaïre Air Boundary
--·- **SPCZ** South Pacific Convergence Zone
   H   Centre of high pressure
  ——   Equator

Figure 2.5 Surface winds

Latitudinal variations in the ITCZ occur as a result of the movement of the overhead sun.

● In June the ITCZ lies further north, whereas in December it lies in the southern hemisphere.
● The seasonal variation in the ITCZ is greatest over large land masses (e.g. Asia).
● By contrast, over the Atlantic and Pacific Oceans its movement is far less.

The word monsoon means reverse and refers to a seasonal reversal of wind direction.

● The monsoon is induced by Asia – the world's largest continent – which causes winds to blow outwards from high pressure in winter but pulls the southern trades into low pressure in the summer.
● The monsoon is therefore influenced by the reversal of land and sea temperatures between Asia and the Pacific during the summer and winter.
● In winter surface temperatures in Asia can be as low as −20°C. By contrast the surrounding oceans have temperatures of 20°C.
● During the summer the land heats up quickly and may reach 40°C. By contrast the sea remains cooler at about 27°C.

● This initiates a land–sea breeze blowing from the cooler sea (high pressure) in summer to the warmer land (low pressure), whereas in winter air flows out of the cold land mass (high pressure) to the warm water (low pressure).

The uneven pattern in Figure 2.5 is the result of seasonal variations in the overhead sun. Summer in the southern hemisphere means that there is a cooling in the northern hemisphere, thereby increasing the temperature differences between polar and equatorial air. Consequently, the high-level westerlies are stronger in the northern hemisphere in winter.

Now test yourself

18 What does the word monsoon mean?

19 In what direction does the monsoon blow in (a) July and (b) January?

Answers on p.214

Tested

Explaining variations in temperature, pressure and winds

Revised

Latitude

Latitude
Areas that are close to the equator receive more heat than areas that are close to the poles. This is due to two reasons:

1 incoming solar radiation (insolation) is concentrated near the equator, but dispersed near the poles.

2 insolation near the poles has to pass through a greater amount of atmosphere and there is more chance of it being reflected back out to space.

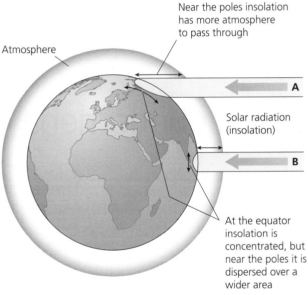

Near the poles insolation has more atmosphere to pass through

Atmosphere

Solar radiation (insolation)

A

B

At the equator insolation is concentrated, but near the poles it is dispersed over a wider area

Source: Nagle, G., *Geography Through Diagrams*, OUP, 1998

Figure 2.6 Latitudinal contrasts in insolation

Distribution of land and sea

There are important differences between the **specific heat capacities** of land and water. Land heats and cools more quickly than water. It takes five times as much heat to raise the temperature of water by 2°C as it does to raise land temperatures.

Water also heats more slowly because:

● it is clear, so the sun's rays penetrate to a greater depth (distributing energy over a larger volume)

● tides and currents cause the heat to be further distributed

Distance from the sea has an important influence on temperature. Water takes up heat and gives it back much more slowly than the land. In winter, in mid-latitudes, sea air is much warmer than the land air, so onshore winds bring heat to the coastal lands. By contrast, during the summer, coastal areas remain much cooler than inland sites. Areas with a coastal influence are termed **maritime** or **oceanic**, whereas inland areas are called **continental**.

> **Specific heat capacity** is the amount of heat needed to raise the temperature of a body by 1°C. There are important differences between the heating and cooling of water.

Sea currents

The effect of ocean currents on temperatures depends upon whether the current is cold or warm. Warm currents from equatorial regions raise the temperatures of polar areas (with the aid of prevailing westerly winds). However, the effect is only noticeable in winter. For example, the North Atlantic Drift raises the winter temperatures of northwest Europe. By contrast, there are other areas that are made colder by ocean currents. Cold currents, such as the Labrador Current off the northeast coast of North America can reduce summer temperature, but only if the wind blows from the sea to the land.

> **Typical mistake**
>
> Some students state that all coastal areas have mild temperatures – it all depends on the temperature of the ocean current.

Factors affecting air movement
Revised

Pressure and wind

The basic cause of air motion is the unequal heating of the Earth's surface. The major equalising factor is the transfer of heat by air movement. Variable heating of the earth causes variations in pressure and this in turn sets the air in motion. There is thus a basic correlation between winds and pressure.

Pressure gradient

The driving force is the **pressure gradient**, i.e. the difference in pressure between any two points. Air blows from high pressure to low pressure (Figure 2.5). Globally, very high-pressure conditions exist over Asia in winter due to the low temperatures. By contrast, the mean sea-level pressure is low over continents in summer. High surface temperatures produce atmospheric expansion and therefore a reduction in air pressure.

> **Now test yourself**
>
> 20 What is meant by the term 'specific heat capacity'?
> 21 State **two** reasons why the poles receive less insolation than the tropics.
>
> **Answers on p.214**
>
> Tested

2.3 Weather processes and phenomena

Moisture in the atmosphere
Revised

Atmospheric moisture exists in all three states – vapour, liquid and solid. Energy is used in the change from one phase to another, for example between a liquid and a gas.

In **evaporation**, water changes from a liquid to a gas, and heat is absorbed. Evaporation depends on three main factors:

- initial humidity of the air – if air is very dry then strong evaporation occurs; if it is saturated then very little occurs
- supply of heat – the hotter the air the more evaporation that takes place
- wind strength – under calm conditions the air becomes saturated rapidly

When **condensation** occurs latent heat locked in the water vapour is released, causing a rise in temperature. Condensation occurs when either (a) enough water vapour is evaporated into an air mass for it to become saturated or (b) when the temperature drops so that the dew point (the temperature at which air is saturated) is reached. The first is relatively rare, the second common. Such cooling occurs in three main ways:

- radiation cooling of the air
- contact cooling of the air when it rests over a cold surface
- adiabatic (expansive) cooling of air when it rises

Condensation requires particles or nuclei onto which the vapour can condense. In the lower atmosphere these are quite common, for example as sea salt, dust and pollution particles. Some of these particles are hygroscopic – they attract water.

When water vapour **freezes**, heat is released. In contrast, heat is absorbed in the process of **sublimation**. When liquid **freezes**, heat is released and temperatures drop. In contrast, when solids **melt**, heat is absorbed and temperatures rise.

Humidity

Absolute humidity refers to the amount of water in the atmosphere. For example, there may be 8 grams of water in a cubic metre of air. **Relative humidity** refers to the water vapour present expressed as a percentage of the maximum amount air at that temperature can hold.

Precipitation

The term 'precipitation' refers to all forms of deposition of moisture from the atmosphere in either solid or liquid states. It includes rain, hail, snow and dew. Because rain is the most common form of precipitation in many areas, the term is sometimes applied to rainfall alone. For any type of precipitation to form, clouds must first be produced.

Now test yourself

22 What happens during evaporation?

23 What happens during condensation?

24 What happens during the process of sublimation?

Answers on p.214

Tested

Expert tip

Saturated air is air with a relative humidity of 100%. As air temperature rises, if there is no increase in water vapour in the air, its relative humidity decreases. As the air is warmed the amount of moisture it can hold increases.

Adiabatic processes (lapse rates)

Revised

Normal or **environmental lapse rate** (ELR) is the actual temperature decline with height – on average, 6°C/km.

Adiabatic cooling and warming in dry (unsaturated) air occurs at a rate of approximately 10°C/km. This is known as the **dry adiabatic lapse rate (DALR)**. Air in which condensation is occurring cools at the lower **saturated adiabatic lapse rate (SALR)** of 4–9°C/km. The rate varies because of the amount of latent heat released. It will be less for warm saturated air (4°C/km) than cold saturated air (9°C/km). However, an average of 5°C/km is generally accepted for the SALR.

Lapse rates can be shown on a temperature/height diagram (Figure 2.7). For example, air temperature of 20°C may have a dew point of 10°C. At first when the air is lifted it cools at the DALR. When it reaches dew point (the temperature at which air is saturated and condensation occurs) it cools at SALR. Saturation level is the same as condensation level. This marks the base of the cloud. Air continues to rise at the SALR until it reaches the same temperature and density as the surrounding air. This marks the top of the cloud development. Thus the changes in lapse rates can be used to show the lower and upper levels of development in a cumulus cloud.

Expert tip

Adiabatic processes relate to the rising and sinking of air. This means that the temperature of the air is changed internally, i.e. without any other influence. It is the rising (expanding and cooling) and sinking (contracting and warming) of air that causes it to change temperature. As air rises in the atmosphere it is cooled and there may be condensation.

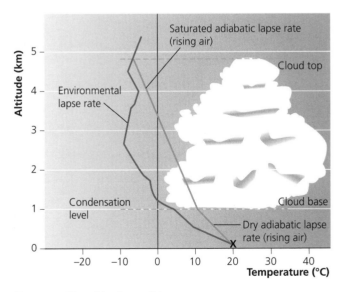

Figure 2.7 Unstable air conditions

Now test yourself

Study Figure 2.7, which shows an environmental lapse rate.

25 At what height will a parcel of air rising from the ground level at X become stable? (Assume the DALR is 10°C/km and the SALR is 5°C/km.)

26 What is the significance of the 'condensation level'?

27 Suggest **three** different causes of the initial uplift of the parcel of air.

Answers on p.214

Tested

Atmospheric stability and instability

Instability occurs when a parcel of air is warmer and therefore less dense than the air above, causing rising and expansion. Uplift and adiabatic cooling of moist air occurs. Air is unstable if ELR > DALR, as in Figure 2.7. Unstable air tends to occur on very hot days when the ground layers are heated considerably.

Stability occurs when ELR < DALR and the SALR (Figure 2.8). If a parcel of air is displaced upwards, it immediately gets cooler and denser and sinks. The only time when stable air can rise is when it is forced to, such as over high ground.

> **Instability** relates to the atmospheric conditions associated with rising air, low-pressure conditions, cloud formation, rain and wind.
>
> **Stability** relates to the atmospheric conditions associated with dry, descending air, which is characterised by calm conditions and relatively clear skies.

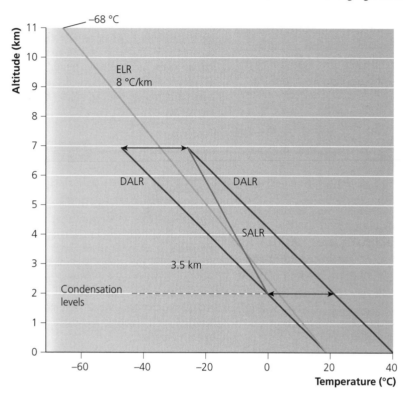

Figure 2.8 Stable air conditions

Conditional instability occurs when ELR lies between SALR and DALR; moist saturated air will rise, whereas dry unsaturated air will sink. Air is therefore stable in respect to the dry rate and would normally sink to its original level. But if air should then become saturated, owing to being forced to rise to higher elevations, it may become warmer than surrounding air and would continue to rise of its own accord.

A very noticeable effect of adiabatic processes is the Fohn effect (Figure 2.9), as demonstrated in the European Alps. Winds approach the mountains as very warm, moist air streams, rise, quickly reach condensation level and therefore are cooling at SALR. At the summit, most moisture is already lost. Hence, on descent winds warm up at DALR. They reach the plains/valleys as hot winds with low relative humidity, and can have a drastic effect by clearing snow very rapidly.

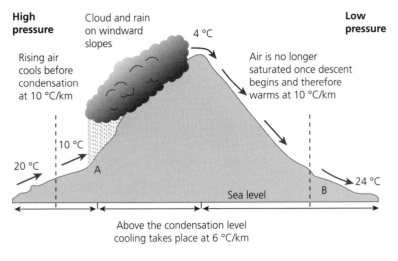
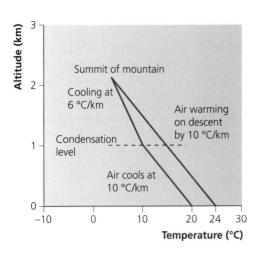

Figure 2.9 Air flow on a mountain — the Fohn effect

Now test yourself
Tested

28 Define the terms instability and stability.

Answer on p.214

Weather phenomena
Revised

Atmospheric stability and instability are closely linked to weather phenomena. Stability means that air does not rise. Stability can lead to the formation of fog, mist and frost. Under clear skies, temperatures may drop enough to form frost. Where there is moisture present, the cooling of air at night may be sufficient to produce mist and fog. Instability, however, produces unstable or rising air, forming clouds and possibly rain.

Clouds

Clouds are formed of millions of tiny water droplets held in suspension. They are classified in a number of ways, the most important being:

● form or shape, such as stratiform (layered) and cumuliform (heaped)
● height, such as low (<2000 m), medium or alto (2000–7000 m) and high (7000–13,000 m)

The important facts to keep in mind are:

● in unstable conditions the dominant form of uplift is convection and this may form cumulus clouds
● where fronts are involved a variety of clouds exist
● relief or topography causes stratiform or cumuliform clouds, depending on stability of the air

Convectional rainfall

When the land becomes very hot it heats the air above it. This air expands and rises. As it rises, cooling and condensation take place. If it continues to rise rain will fall. This is very common in tropical areas. In temperate areas, convectional rain is more common in summer.

Frontal or cyclonic rainfall

Frontal rain occurs when warm air meets cold air. The warm air, being lighter and less dense, is forced to rise over the cold, denser air. As it rises it cools, condenses and forms rain. It is most common in middle and high latitudes, where warm tropical air and cold polar air converge.

Relief or orographic rainfall

Air may be forced to rise over a barrier such as a mountain. As it rises it cools, condenses and forms rain. There is often a rain shadow effect whereby the leeward slope receives a relatively small amount of rain. Altitude is important especially on a local scale. In general, there are increases of precipitation up to about 2 km. Above this level rainfall decreases because of the air temperature being so low.

Hail

Hail is made up of alternate shells of clear and opaque ice, formed by raindrops being carried up and down in vertical air currents in large cumulonimbus clouds. Freezing and partial melting may occur several times before the pellet is large enough to escape from the cloud. As a raindrop is carried high up in the cloud it freezes. As the hailstone falls, the outer layer may be melted but can freeze again with further uplift. The process can occur many times before the hail finally falls to ground, that is, when its weight is great enough to overcome the strong updraughts of air.

Snow

Snow is frozen precipitation. Snow crystals form when the temperature is below freezing point and water vapour is converted into a solid. However, very cold air contains a limited amount of moisture, so the heaviest snowfalls tend to occur when warm moist air is forced over very high mountains or when warm moist air comes into contact with very cold air at a front.

Frost

Frost is a deposit of fine ice crystals onto the ground or vegetation. It occurs on cloud-free nights when there has been radiation cooling to below freezing point. Water vapour condenses directly onto these surfaces by sublimation.

Dew

Dew is the direct deposition of water droplets onto the surface and vegetation. It occurs in clear, calm anticyclonic conditions (stability) where there is rapid radiation cooling by night. The temperature reaches dew point, and further cooling causes condensation and direct precipitation onto the ground and vegetation.

Fog

Fog is cloud at ground level. **Radiation fog** is formed in low-lying areas during calm weather, especially during spring and autumn. The surface of the ground, cooled rapidly at night by radiation, cools the air immediately above it. This air then flows into hollows by gravity and is cooled to **dew point**, causing condensation. Ideal conditions include a surface layer of moist air and clear skies to allow maximum **radiation cooling** to occur quickly.

As the sun rises, radiation fog clears away. Under cold anticyclonic conditions in late autumn and winter, fog may be thicker and more persistent, and around large towns smog may develop under an inversion layer.

Fog commonly occurs over the sea in autumn and spring because the contrast in temperature between land and sea is significant. **Advection fog** is formed when warm moist air flows horizontally over cooler land or sea.

> **Typical mistake**
>
> Some students think that fog is common all year round. It is more common in temperate areas in spring and autumn. In summer, the sea is cooler than the land so air is not cooled when it blows onto the land, while in winter there are more low pressure systems, causing higher winds and mixing the air.

Now test yourself

Tested ☐

29 Distinguish between radiation fog and advection fog.

30 Under which atmospheric conditions (stability or instability) do mist and fog form?

Answers on p.214

2.4 The human impact

Global warming

The enhanced greenhouse effect

The greenhouse effect is both natural and good – without it there would be no human life on Earth. On the other hand, there are concerns about the **enhanced greenhouse effect**.

The enhanced greenhouse effect is a build up of certain **greenhouse gases** as a result of human activity. Studies of cores taken from ice packs in Antarctica and Greenland show that the level of CO_2 between 10,000 years ago and the mid-nineteenth century was stable at about 270 ppm. By 1957 the concentration of CO_2 atmosphere was 315 ppm. It has since risen to about 360 ppm and is expected to reach 600 ppm by 2050. The increase is due to human activities – primarily the burning of fossil fuels (coal, oil and natural gas) and deforestation. Deforestation of the tropical rainforest also increases atmospheric CO_2 levels because it removes the trees that convert CO_2 into oxygen.

> **Greenhouse gases**, such as water vapour, CO_2, methane, ozone, nitrous oxides and chlorofluorocarbons (CFCs), like the glass on a greenhouse, allow short-wave radiation from the Sun to pass through, but they trap outgoing long-wave radiation, thereby raising the temperature of the lower atmosphere.

Table 2.2 Properties of key greenhouse gases

	Average atmosphereic concentration (ppmv)	Rate of change (% per annum)	Direct global warming potential (GWP)	Lifetime (years)	Type of indirect effect
CO_2	355	0.5	1	120	None
Methane	1.72	0.6–0.75	11	10.5	Positive
Nitrous oxide	0.31	0.2–0.3	270	132	Uncertain
CFC-11	0.000255	4	3400	55	Negative
CFC-12	0.000453	4	7100	116	Negative

Climate change

Climate change is a very complex issue for a number of reasons:

- It involves interactions between the atmosphere, oceans and land masses.
- It includes natural as well as anthropogenic forces.
- There are feedback mechanisms, not all of which are fully understood.
- Many of the processes are long term and so the impact of changes may not yet have occurred.

The effects of increased global temperature

The effects of global warming are very varied. Much depends on the scale of the changes. For example, some impacts could include:

- a rise in sea levels, causing flooding in low-lying areas such as the Netherlands, Egypt and Bangladesh – up to **200 million** people could be displaced
- an increase in storm activity, such as more frequent and intense hurricanes (owing to more atmospheric energy)
- **4 billion** people suffering from water shortage if temperatures rise by **2°C**
- **35%** drop in crop yields across Africa and the Middle East if temperatures rise by **3°C**
- **200 million** more people could be exposed to hunger if world temperatures rise by **2°C**, **550 million** if temperatures rise by **3°C**
- extinction of up to **40%** of species of wildlife if temperatures rise by **2°C**

> **Expert tip**
>
> There are many causes of global climate change. Natural causes include:
> - variations in the Earth's orbit around the Sun
> - variations in the tilt of the Earth's axis
> - variations in solar output (sunspot activity)
> - changes in the amount of dust in the atmosphere (partly due to volcanic activity)
> - changes in the Earth's ocean currents as a result of continental drift
>
> All of these have helped cause climate change, and may still be doing so, despite anthropogenic (human-generated) forces.

The Stern Report (2006) is a report by Sir Nicholas Stern analysing the financial implications of climate change. The report has a simple message:

- Climate change is fundamentally altering the planet.
- The risks of inaction are high.
- Time is running out.

The effects of climate change vary with the degree of temperature change (Figure 2.10). According to the Stern Report, global warming could deliver an economic blow of between 5% and 20% of GDP to world economies. Dealing with the problem, by comparison, will cost just 1% of GDP, equivalent to £184 billion.

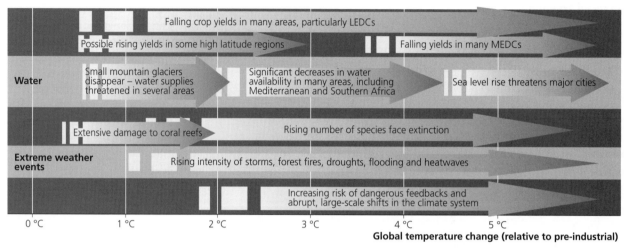

Figure 2.10 Projected impacts of climate change, according to the Stern Report

Now test yourself

Figure 2.10 shows some of the projected impacts related to global warming.

31 Describe the potential changes of a 4°C rise in temperature.

32 Explain why there is an increased risk of hazards in coastal cities.

33 Outline the ways in which it is possible to manage the impacts of global warming.

Answers on p.214

> **Typical mistake**
>
> The Kyoto Protocol (1997) did not cover all countries. It gave all MEDCs legally binding targets for cuts in emissions from the 1990 level by 2008–2012. This was extended to 2017. The EU agreed to cut emissions by 8% and Japan by 7%.

Urban climates

Urban climates occur as a result of extra sources of heat released from industrial, commercial and residential buildings as well as from vehicles. In addition, concrete, glass, bricks and tarmac all act very differently from soil and vegetation. For example, the albedo (reflectivity) of tarmac is about 5–10% while that of concrete is 17–27%. In contrast, that of grass is 20–30%. Some of these materials – notably dark bricks – absorb large quantities of heat and release them slowly by night. In addition, the release of pollutants helps trap radiation in urban areas. Consequently, urban microclimates can be very different from rural ones. Greater amounts of dust mean increasing concentration of hygroscopic particles. There is less water vapour, but more CO_2 and higher proportions of noxious fumes owing to combustion of imported fuels. Discharge of waste gases by industry is also significant.

Urban heat budgets differ from rural ones. By day the major source of heat is solar energy; in urban areas brick, concrete and stone have high heat capacities. The extensive surfaces of these materials in urban areas allow a greater area to be heated.

> **Urban climates** refer to the changes in temperature, humidity, wind patterns, precipitation and air pressure that are noticeable over large urban areas during high-pressure conditions.

In urban areas there is relative lack of **moisture**. This is due to:

● lack of vegetation
● high drainage density (sewers and drains), which remove water

Thus, there are decreases in relative humidity in inner cities due to the lack of available moisture and higher temperatures there.

Nevertheless, there are more intense **storms**, particularly during hot summer evenings and nights, owing to greater instability and stronger convection above build-up areas. There is a higher incidence of **thunder**, but less **snow fall**.

At night the ground radiates heat and cools; in urban areas the release of heat by buildings offsets the cooling process, in addition, some industries, commercial activities and transport networks continue to release heat throughout the night.

There is higher incidence of thicker **cloud cover** in summer and radiation fogs or smogs in winter because of increased convection and air pollution respectively. The concentration of hygroscopic particles accelerates the onset of condensation.

Expert tip

The contrasts between urban and rural areas are greatest under calm, high-pressure conditions. The typical heat profile of an urban heat island shows the maximum at the city centre, a plateau across the suburbs and a temperature cliff between the suburban and rural area. Small-scale variations within the urban heat island occur, with the distribution of industries, open space, rivers, canals and so on.

Table 2.3 Average changes in climate caused by urbanisation

Factor		Comparison with rural environments
Radiation	Global	2–10% less
	Ultraviolet, winter	30% less
	Ultraviolet, summer	5% less
	Sunshine duration	5–15% less
Temperature	Annual mean	1°C more
	Sunshine days	2–6°C more
	Greatest difference at night	11°C more
	Winter maximum	1.5°C more
	Frost free season	2–3 weeks more
Wind speed	Annual mean	10–20% less
	Gusts	10–20% less
	Calms	5–20% more
Relative humidity	Winter	2% less
	Summer	8–10% less
Precipitation	Total	5–30% more
	Number of rain days	10% more
	Snow days	14% less
Cloudiness	Cover	5–10% more
	Fog, winter	100% more
	Fog, summer	30% more
	Condensation nuclei	10 times more
	Gases	5–25 times more

Now test yourself

34 Describe the main differences between the climates of urban areas and those of their surrounding rural areas.

35 What is meant by the urban heat island?

36 Describe **one** effect that atmospheric pollution may have on urban climates.

37 Why are microclimates, such as urban heat islands, best observed during high-pressure (anticyclonic) weather conditions?

Answers on p.214

Tested

Daytime temperatures in rural areas are, on average, 0.6°C warmer. This **urban heat island** effect is noticeable, especially by dawn during anticyclonic conditions. The effect is caused by a number of factors:

● heat produced by human activity
● buildings having a high thermal capacity in comparison with rural areas – up to six times greater than agricultural land
● fewer bodies of open water (therefore less evaporation) and fewer plants (therefore less transpiration)

- the composition of the atmosphere, involving the blanketing effect of smog, smoke or haze
- less thermal energy required for evaporation and evapotranspiration due to the surface character, rapid drainage and generally lower wind speeds

Air flow over an urban area is disrupted – winds are slow and deflected over buildings. Large buildings can produce eddying. Severe gusting and turbulence around tall buildings causes strong local pressure gradients from windward to leeward walls. Deep, narrow streets are much calmer unless aligned with prevailing winds to funnel flows along them – the 'canyon effect'.

The nature of urban climates is changing, however. With the decline in coal as a source of energy there is less SO_2 pollution, hence fewer hygroscopic nuclei and so less fog.

Exam-style questions

Section A

1 (a) Explain the meaning of the terms insolation and albedo. [2]

(b) How would insolation and albedo vary between a polar ice cap and a tropical rainforest? [2]

2 Explain why a coastal area may have a different temperature from a continental area of the same latitude. [3]

3 Using examples, explain how aspect can affect temperature. [3]

Section B

1 (a) Define the terms dry adiabatic lapse rate (DALR) and saturated adiabatic lapse rate (SALR). [4]

(b) Briefly explain how conditional instability occurs. [3]

2 Using an annotated diagram, show how an understanding of lapse rates can help explain cloud formation. [8]

3 Explain how atmospheric stability, instability and conditional instability lead to different weather conditions. [10]

Exam ready

3 Rocks and weathering

3.1 Elementary plate tectonics

The structure of the Earth
Revised

The theory of plate tectonics states that the Earth is made up of a number of layers. On the outside there is a very thin crust, and underneath is a mantle, which makes up 82% of the volume of the Earth. Deeper still is a very dense and very hot core.

Close to the surface rocks are mainly solid and brittle. This upper surface layer, known as the **lithosphere**, includes the **crust** and the upper **mantle**, and is about 70 km deep. The Earth's crust is commonly divided up into two main types – **continental crust** and **oceanic crust** (Table 3.1).

Table 3.1 A comparison of oceanic crust and continental crust

Characteristics	Continental crust	Oceanic crust
Thickness	35–70 km on average	6–10 km on average
Age of rocks	Very old, mainly over 1500 million years	Very young, mainly under 200 million years
Colour and density of rocks	Light in colour; lighter, with an average density of 2.6	Dark in colour; heavier, with an average density of 3.0
Minerals	Silica, aluminium and oxygen	Silica, iron and magnesium
Nature of rocks	Numerous types; many contain silica and oxygen; granitic is the most common	Few types; mainly basaltic

Plate boundaries
The zone of earthquakes around the world has helped to define six major plates and a number of minor plates (Figure 3.1).

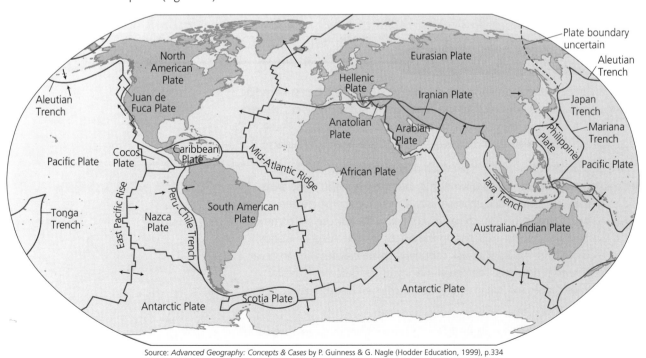

Source: *Advanced Geography: Concepts & Cases* by P. Guinness & G. Nagle (Hodder Education, 1999), p.334

Figure 3.1 Plate boundaries

The movement of plates and continental drift

There are three main theories about movement.

- The convection current theory states that huge convection currents occur in the Earth's interior. Magma rises through the core to the surface and then spreads out at mid-ocean ridges. The cause of the movement is radioactive decay in the core.
- The dragging theory states that plates are dragged or subducted by their oldest edges, which have become cold and heavy. Plates are hot at the mid-ocean ridge but cool as they move away. Complete cooling takes about a million years. As cold plates descend at the trenches pressure causes the rock to change and become heavier.
- A hotspot is a plume of lava that rises vertically through the mantle. Most are found near plate margins and they may be responsible for the original rifting of the crust. However, the world's most abundant source of lava, the Hawaiian hotspot, is not on the plate margin. Hotspots can cause movement – the outward flow of viscous rock from the centre may create a drag force on the plates and cause them to move.

Sea-floor spreading

In the early 1960s Dietz and Hess suggested that continents moved in response to the growth of oceanic crust between them. Confirmation of sea-floor spreading came with the discovery by Vine and Matthews that magnetic anomalies across the Mid-Atlantic Ridge were symmetrical on either side of the ridge axis. The only acceptable explanation for these magnetic anomalies was in terms of sea-floor spreading and the creation of new oceanic crust. When lava cools on the sea floor, magnetic grains in the rock acquire the direction of the Earth's magnetic field at the time of cooling. This is known as paleomagnetism.

- Slow-spreading ridges, such as the Mid-Atlantic Ridge, have a pronounced rift down the centre. Slow-spreading ridges are fed by small and discontinuous magma chambers, thereby allowing for the eruption of a comparatively wide range of basalt types.
- Fast-spreading ridges, such as the East Pacific Rise, lack the central rift and have a smooth topography. In addition, spreading rates have not remained constant through time. Fast-spreading ridges have large, continuous magma chambers that generate comparatively similar magmas. Because of the higher rates of magma discharge, sheet lavas are more common.

> **Continental drift** is the movement of continents across the surface of the globe.
>
> **Sea-floor spreading** occurs at constructive margins, where ocean floors grow as plates move apart.

Expert tip

The reason why the ridges are elevated above the ocean floor is that they consist of rock that is hotter and less dense than the older, colder plate.

Typical mistake

Spreading rates are not the same throughout the mid-ocean ridge system but vary considerably from a few millimetres per year in the Gulf of Aden to 1 cm per year in the North Atlantic near Iceland and 6 cm per year for the East Pacific Rise.

Now test yourself

1 Outline the main differences between continental crust and oceanic crust.
2 Name the six major plates shown in Figure 3.1.3
3 Briefly explain what is meant by (a) paleomagnetism and (b) sea-floor spreading.

Answers on p.214

Tested

Processes at plate boundaries

Revised

The boundaries between plates can be divided into three main types: spreading (constructive), colliding (destructive) and conservative. Spreading ridges, where new crust is formed, are mostly in the middle of oceans (Figure 3.2a). These ridges are zones of shallow earthquakes (less than 50 km below the surface).

Where two plates collide a deep-sea trench may be formed where one of the plates is **subducted** (forced downwards) into the mantle. In these areas, fold mountains are formed and chains of island arcs may also be created (Figure 3.2b). Deep earthquakes, up to 700 km below the surface, are common. Good examples include the trenches off the Andes and the Aleutian Islands that stretch out from Alaska. If a thick continental plate collides with an ocean plate a deep trench develops. The partial melting of the descending ocean plate causes volcanoes to form in an arc-shaped chain of islands, such as in the Caribbean.

Along some plate boundaries plates slide past one another to create a transform fault (fault zone) without colliding or separating (Figure 3.2c). Again these are associated with shallow earthquakes, such as the San Andreas Fault in California.

Where continents embedded in the plates collide with each other there is no subduction, but crushing and folding can create young fold mountains such as the Himalayas (Figure 3.2d).

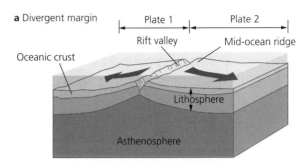

a Divergent margin

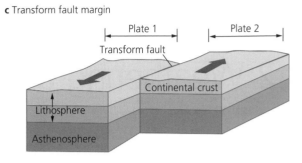

c Transform fault margin

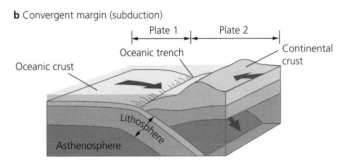

b Convergent margin (subduction)

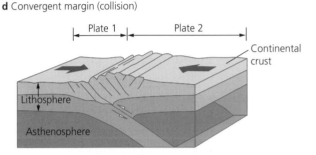

d Convergent margin (collision)

Source: *Advanced Geography: Concepts & Cases* by P. Guinness & G. Nagle (Hodder Education, 1999), p.334

Figure 3.2 Types of plate boundary

Mountain building

Plate tectonics is associated with mountain building. Linear chains are associated with convergent plate boundaries, and formed on land. Where an ocean plate meets a continental plate, the lighter, less dense continental plate may be folded and buckled into fold mountains, such as the Andes. Where two continental plates meet, both may be folded and buckled, as in the case of the Himalayas, formed by the collision of the Eurasian and Indian plates.

Ocean ridges

The longest linear, uplifted features of the Earth's surface are to be found in the oceans. They are giant submarine mountain chains with a total length of more than 60,000 km, are 1000–4000 km wide, and have crests that rise 2–3 km above the surrounding ocean basins, which are 5 km deep. The average depth of water over their crests is thus about 2500 m.

Ocean trenches

Subduction zones form where an oceanic lithospheric plate collides with another plate – **continental** or **oceanic**. Subducted (lithospheric) oceanic crust remains cooler, and therefore denser, than the surrounding mantle for millions of years, so once initiated subduction carries on, driven in part by the weight of the subducting crust. As the Earth has not grown significantly in size – not enough to accommodate the new crustal material created at mid-ocean ridges – the amount of subduction roughly balances the amount of production at the constructive plate margins.

At the subduction zone, deep-sea **trenches** are found. Deep-sea trenches are long, narrow depressions in the ocean floor with depths of over 6000 m and up to 11,000 m. Trenches are found adjacent to land areas and associated with island arcs. They are more numerous in the Pacific Ocean. The trench is usually asymmetric, with the steep side towards the land mass. Where a trench occurs off continental margins, the turbidites (sediments) from the slope are trapped, forming a hadal plain on the floor of the trench.

Now test yourself

4 What processes happen at (a) a mid-ocean ridge and (b) a subduction zone?

5 With which types of plate boundary is volcanic activity associated?

Answers on p.214

Tested

Expert tip

Mountain building is often associated with crustal thickening, deformation and volcanic activity, although in the case of the Himalayas, volcanic activity is relatively unimportant.

Expert tip

Subduction zones dip mostly at angles between 30° and 70°, but individual subduction zones dip more steeply with depth. The dip of the slab is related inversely to the velocity of convergence at the trench, and is a function of the time since the initiation of subduction. The older the crust the steeper it dips. Because the down-going slab of lithosphere is heavier than the plastic asthenosphere below, it tends to sink passively.

Island arcs

Island arcs are a series of volcanic islands, formed in an arc-shape, as in the Caribbean when oceanic lithosphere is subducted beneath oceanic lithosphere.

3.2 Weathering and rocks

Weathering works in different ways:

- **Physical** weathering produces smaller, angular fragments of the same rock, such as scree.
- **Chemical** weathering creates altered rock substances, such as kaolinite (china clay) from granite.
- A third type, **biological** weathering, has been identified, whereby plants and animals chemically alter rocks and physically break them through their growth and movement.

Now test yourself

6 Describe the main features of an island arc system.

7 Briefly explain how island arcs are formed.

Answers on p.214

Tested

Weathering is the decomposition and disintegration of rocks in situ.

Expert tip

It is important to note that these processes are *interrelated* rather than operating in isolation.

Physical weathering

Revised

Physical weathering operates at or near the Earth's surface, where temperature changes are most frequent.

- **Freeze–thaw** (also called frost shattering) occurs when water in joints and cracks freezes at 0°C. It expands by about 10% and exerts pressure up to a maximum of 2100 kg/cm^2 at −22°C. Freeze–thaw is most effective in environments where moisture is plentiful and there are frequent fluctuations above and below freezing point. Hence it is most effective in periglacial and alpine regions.

- Heating and cooling of rocks can also lead to weathering. Different minerals expand and contract at different temperatures. This can cause **granular disintegration** in rocks composed of different minerals, for example granite contains quartz, feldspar and mica. In contrast, where the rock consists of a single mineral, **block disintegration** is more likely.

- **Disintegration** is found in hot desert areas where there is a large diurnal temperature range. In many desert areas daytime temperatures exceed 40°C whereas night-time ones are little above freezing. Rocks heat up by day and contract by night. As rock is a poor conductor of heat, stresses occur only in the outer layers. This causes peeling or **exfoliation** to occur. Moisture is essential for this to happen.

- Cycles of **wetting and drying** can also lead to the breakdown of rock. Salt crystals, for example, expand when water is added to them. Wetting and drying are particularly effective on shale rocks.

- **Salt crystallisation** causes the decomposition of rock by solutions of salt:
  - In areas where temperatures fluctuate around 26–28°C sodium sulfate (Na_2SO_4) and sodium carbonate (Na_2CO_3) expand by about 300%. This creates pressure on joints, forcing them to crack.
  - When water evaporates, salt crystals may be left behind. As the temperature rises, the salts expand and exert pressure on rock.

  This mechanism is frequent in hot desert regions where low rainfall and high temperatures cause salts to accumulate just below the surface.

- **Pressure release** is the process whereby overlying rocks are removed by erosion. This causes underlying rocks to expand and fracture parallel to the surface. The removal of a great weight, such as a glacier, has the same effect.

Now test yourself

8 Define physical weathering.

9 What are the factors that make freeze–thaw weathering effective?

10 Describe the process of exfoliation. Why is it characteristic of hot desert environments?

Answers on p.214

Tested

Chemical weathering

Water is the key medium for chemical weathering. Unlike mechanical weathering, chemical weathering is most effective sub-surface because percolating water has gained organic acids from the soil and vegetation.

- Acidic water helps to break down rocks such as chalk, limestone and granite. The amount of water is important as it removes weathered products by solution.
- **Hydrolysis** occurs on rocks with orthoclase feldspar – notably granite.
- Feldspar reacts with acid water and forms **kaolin**, silicic acid and potassium hydroxyl. The acid and hydroxyl are removed in the solution leaving **kaolin** behind as the end product. Other minerals in the granite, such as quartz and mica, remain in the kaolin.
- **Hydration** is the process whereby certain minerals absorb water, expand and change. For example, anhydrite is changed to gypsum. Although it is often classified as a type of chemical weathering, mechanical stresses occur as well. When anhydrite absorbs water to become gypsum it expands by about 0.5%. Shales and mudstones increase in volume by around 100% when clay minerals absorb water.
- **Carbonation–solution** occurs in rocks with calcium carbonate, such as chalk and limestone. Rainfall combines with dissolved carbon dioxide or organic acid to form a weak carbonic acid.
- Calcium carbonate (calcite) reacts with acid water and forms **calcium bicarbonate**, which is soluble and removed in **solution** by percolating water.
- The effectiveness of solution is related to the pH of the water.
- **Oxidation** occurs when iron compounds react with oxygen to produce a reddish brown coating. In this way dissolved oxygen in the soil or the atmosphere affects iron minerals. Oxidation is most common in areas that are well drained. FeO is oxidised to Fe_2O_3. This is soluble only under extreme acidity (pH <3.0). Hence it remains in many soils and rocks, especially in tropical areas, giving a red colour.

Organic action can help weather rocks. Acids derived from the decomposition of vegetation are termed **humic acids**. In addition, bacterial activity and the respiration of plant roots raise CO_2 levels in the soil, thereby aiding solution. **Chelation** is the process in which plant roots can absorb relatively insoluble minerals. This occurs because the roots are surrounded by a concentration of hydrogen ions, which can exchange with cations in adjacent minerals.

> **Typical mistake**
>
> **Reduction** of ferric iron to ferrous iron allows iron oxides to be removed from solution. This typically occurs in waterlogged, marshy areas and forms blue-grey clays associated with anaerobic (oxygen-deficient) conditions. Many students fail to mention the anaerobic conditions associated with waterlogged areas.

> **Now test yourself**
>
> 11 Compare the character of rocks affected by mechanical weathering with those affected by chemical weathering.
>
> **Answer on p.214**
>
> Tested

Controls of weathering

Climate

In the simplest terms, the type and rate of weathering vary with climate (Figure 3.3). But it is very difficult to isolate the exact relationship, at any scale, between climate type and rate of process.

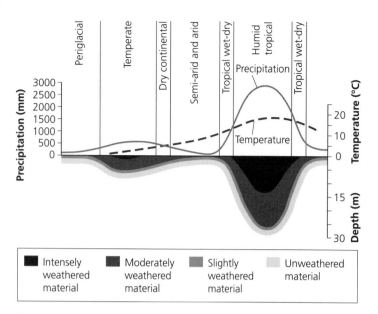

Figure 3.3 Depth of weathering profile and climate

- Peltier's diagram (1950) shows how weathering is related to moisture availability and average annual temperature (Figure 3.4).
- The efficiency of freeze–thaw, salt crystallisation and insolation weathering is influenced by critical temperature changes, frequency of cycles, and diurnal and seasonal variations in temperature.

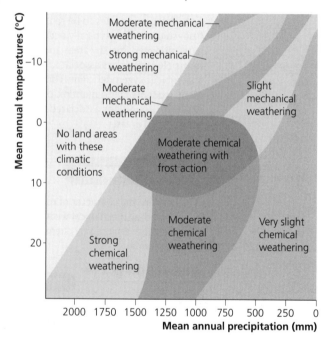

Figure 3.4 Peltier's diagram, showing variations of chemical and mechanical weathering with climate

Geology

Rock type and rock structure influence the rate and type of weathering in many ways depending on:

- chemical composition
- the nature of cements in sedimentary rock
- joints and bedding planes

For example, **limestone** consists of calcium carbonate and is therefore susceptible to carbonation–solution. By contrast, **granite** is prone to hydrolysis because of the presence of feldspar.

> **Typical mistake**
>
> In general, frost shattering increases as the number of freeze–thaw cycles increases rather than according to average temperatures. By contrast, chemical weathering increases with moisture and heat.

> **Expert tip**
>
> According to **Van't Hoff's law**, the rate of chemical weathering increases 2–3 times for every increase of temperature of 10°C (up to a maximum temperature of 60°C).

In **sedimentary** rocks, the nature of the cement is crucial. Iron-oxide-based cements are prone to oxidation whereas quartz cements are very resistant.

The effect of **rock structure** varies from large-scale folding and faulting to localised patterns of joints and bedding planes. Joint patterns exert a strong control on water movement. These act as lines of weakness, thereby creating **differential resistance** within the same rock type.

Similarly, **grain size** influences the speed with which rocks weather. Coarse-grained rocks weather quickly owing to large empty spaces and high permeability. On the other hand, fine-grained rocks offer a greater surface area for weathering and may be highly susceptible to weathering.

The importance of individual **minerals** was stressed by Goldich in 1938. Rocks formed of resistant minerals, such as quartz, muscovite and feldspar in granite, will resist weathering. By contrast, rocks formed of weaker minerals will weather rapidly.

Vegetation
The presence of vegetation can increase weathering through the release of organic acids and through the growth of root systems. The solution of limestone is greater under soil than on bare rock due to the extra organic acids released by the vegetation into the soil.

Relief
- On very steep slopes weathered material may be removed quickly but scree slopes can develop, which protect the rock-face from further weathering.
- On very flat slopes weathered material will not be removed and fresh rock faces will not be exposed.
- On intermediate slopes some weathered material will be removed, exposing fresh rock faces.

Granite
Revised

Granite is an igneous, crystalline rock. It has great physical strength and is very resistant to erosion. There are many types of granite but all contain quartz, mica and feldspar. These are resistant minerals. The main processes of weathering that occurs on granite are freeze–thaw and hydrolysis.

Characteristic granite landscapes include exposed large-scale **batholiths**, which form mountains.

The formation of tors
Tors are isolated masses of bare rock. They can be up to 20 m high. Some of the boulders of the mass are attached to part of the bedrock. Others merely rest on the top.

Linton (1955) argued that the well-developed jointing system (of irregular spacing) was chemically weathered. This occurred under warm, humid conditions during the Tertiary era. Decomposition was most rapid along joint planes. Where the distance between the joint planes was largest, masses of granite remained relatively unweathered. These corestones were essentially embryonic tors. Subsequent denudation has removed the residue of weathering, leaving the unweathered blocks standing out as tors.

An alternative theory relates tor formation to frost action under periglacial conditions. This led to the removal of the more closely jointed portions of the rock. Intense frost shattering under periglacial conditions, followed by removal of material by solifluction, removed the finer material and left the tors standing.

A **tor** is an isolated mass of rock, often granite, left standing on a hilltop after the surrounding rock has been broken down and removed.

Expert tip

Tors are a good example of **equifinality**. This means that different processes can produce the same end result.

Limestone scenery

Limestone scenery is unique on account of its:

- permeability
- solubility in rain and ground-water

Limestone consists of mainly calcium carbonate. Because of their permeability, limestone areas are often dry on the surface and are known as Karst landscapes.

Carboniferous limestone has a distinctive bedding plane and joint pattern, described as **massively jointed**. These features act as weaknesses allowing water to percolate into the rock and dissolve it.

One of the main processes affecting limestone is carbonation–solution. The process is reversible, so under certain conditions calcium carbonate can be deposited in the form of **speleothems** (cave deposits such as **stalactites** and **stalagmites**) and **tufa** (calcium deposits around springs). Limestone is also affected by freeze–thaw, fluvial erosion, glacial erosion and mass movements.

Surface features
As the joints and cracks are attacked and enlarged over thousands of years, the limestone's permeability increases. **Clints** and **grikes** develop on the surface of the exposed limestone. Large areas of bare exposed limestone are known as **limestone pavements**.

Depressions can range from small-scale **swallow holes** (or **sinks**) to large **dolines** up to 30 m in diameter. These are caused by the solution of limestone but can also be formed by the enlargement of a grike system, by carbonation or fluvial activity, or by the collapse of a cavern. Rivers can disappear into these holes, hence the term 'sink'. **Resurgent streams** arise when the limestone is underlain by an impermeable rock, such as clay.

> **Expert tip**
>
> **Stalactites** develop from the top of the cave, whereas **stalagmites** are formed on the base of the cave.

> **Now test yourself**
>
> 16 What are the two main theories about tor formation?
> 17 What is equifinality?
> 18 What are the main processes affecting limestone?
> 19 Explain the formation of swallow holes.
>
> **Answers on p.215**
>
> Tested

3.3 Slope processes and development

Slope development

Rock type
Geological structure is an important influence on slope development. This includes faults, angle of dip and vulcanicity:

- Faulting can produce steep valley sides, as in a rift valley.
- Folding can produce either steep or gentle slopes depending on the angle of dip.
- Vulcanicity produces intrusions of resistant igneous rock. For example, Great Whin Sill is a harder and more resistant rock than the surrounding dolomite, and so has produced a steep slope.

Rock type and character influence whether a rock is affected by weathering, and to what extent it can resist the downslope force of mass movement. Resistance is largely physical. Regular jointing can increase the risk of movement, as well as the amount of water that enters a rock.

Climate
Many slopes are shaped by **climate**, which affects the types and rates of processes that operate in a region and when they occur:

- In arid regions, slopes are jagged or straight due to mechanical weathering and sheetwash.

> A **slope** is an inclined surface (hillslope). It can also refer to the angle of inclination.

> **Expert tip**
>
> Slopes that comprise many different types of rocks are often more vulnerable to landslides due to differential erosion. The softer rocks get worn away and can lead to the undermining of harder rocks.

- In humid areas, slopes are frequently rounder, due to chemical weathering, soil creep and fluvial transport.
- In the humid tropics accelerated chemical weathering dominates. This is due to the hot, wet conditions and the availability of organic acids. Deep clays are produced, favouring low slope angles.

Regolith

Regolith is the superficial and unconsolidated material found at the Earth's surface. It includes soil, scree, weathered bedrock and deposited material.

- Its unconsolidated nature makes it prone to downslope movement. The extra weight of a deep regolith will increase the likelihood of instability.
- Clay-rich regoliths are particularly unstable because of their ability to retain water.

Soil

Soil structure and texture will largely determine how much water it can hold. Clay soils can hold more water than sandy soils. A deep clay on a slope where vegetation has been removed will offer very little resistance to mass movement.

Aspect

Aspect refers to the direction in which a slope faces. In some areas, past climatic conditions varied depending on the direction a slope faced.

Vegetation

Vegetation can decrease runoff through the interception and storage of moisture. Deforested slopes are frequently exposed to intense erosion and gulleying. However, vegetation can also increase the chance of major landslips.

Mass movements

Revised

Mass movements (Figure 3.5) include:

- very slow movements, such as soil creep
- fast movements, such as avalanches
- dry movements, such as rock falls
- fluid movements such as mud flows

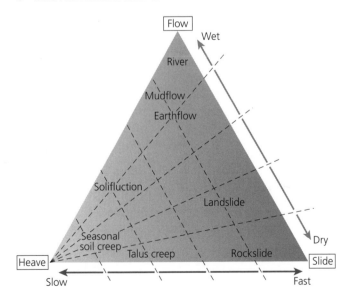

Figure 3.5 A classification of mass movements

Causes of mass movements

The likelihood of a slope failing can be expressed by its safety factor. This is the relative strength or resistance of the slope, compared with the force that is trying to move it. The most important factors that determine movement are gravity, slope angle and pore pressure.

Gravity has two effects:

● It acts to move the material downslope (a slide component).
● It acts to stick the particle to the slope (a stick component).

The downslope movement is proportional to the weight of the particle and slope angle. Water lubricates particles and in some cases fills the spaces between the particles, which forces them apart under pressure.

Types of mass movement

Heave

Heave or **creep** is a slow, small-scale process, which occurs mostly in winter. It is one of the most important slope process in environments where flows and slides are not common. **Talus creep** is the slow movement of fragments on a scree slope.

Individual soil particles are pushed or heaved to the surface by (a) wetting, (b) heating or (c) freezing of water. About 75% of the soil creep movement is induced by moisture changes and associated volume change. Freeze–thaw and normal temperature-controlled expansion and contraction are also important in periglacial and tropical climates. Particles move at right-angles to the surface as it is the path of least resistance. They then fall under the influence of gravity when the particles have dried or cooled, or when the water has thawed. Net movement is downslope. Heave forms terracettes.

Falls

Falls occur on steep slopes (>40°), especially bare rock faces where joints are exposed. The initial cause of the fall may be weathering, such as freeze–thaw or disintegration, or erosion prising open lines of weakness. Once the rocks are detached they fall under the influence of gravity (Figure 3.6). If the fall is short it produces a relatively straight scree. If it is long, it forms a concave scree. Falls are significant in causing the retreat of steep rock faces and in providing debris for scree slopes and talus slopes.

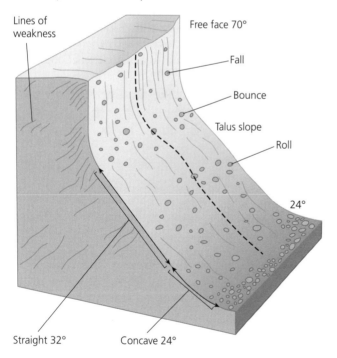

Figure 3.6 Falls

Slides

Slides occur when an entire mass of material moves along a slip plane. They include:

● rockslides and landslides of any material, rock, or regolith
● rotational slides, which produce a series of massive steps or terraces

Now test yourself

21 State **one** difference between a rockslide and a mudflow.
22 Define the term mass movement.
23 Suggest how mass movements can be classified.
24 Define the terms shear strength and shear stress.

Answers on p.215

Tested

Slides commonly occur where there is a combination of weak rocks, steep slopes and active undercutting. They are often caused by a change in the water content of a slope or by very cold conditions. As the mass moves along the slip plane it tends to retain its shape and structure until it impacts at the bottom of a slope (Figure 3.7). Slides range from small-scale slides close to roads, to large-scale movements killing thousands of people.

Slip planes occur:

- at the junction of two layers
- at a fault line
- where there is a joint
- along a bedding plane
- at the point beneath the surface where the shear stress becomes greater than the shear strength

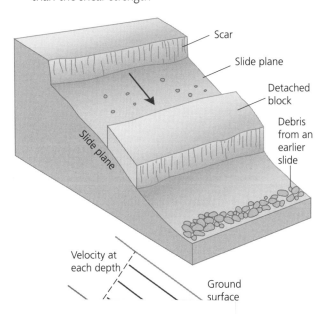

Scar
Slide plane
Detached block
Debris from an earlier slide
Slide plane
Velocity at each depth
Ground surface

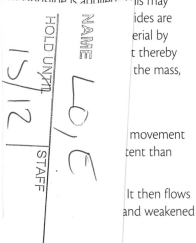

...ncy to move downslope. They will do ... the resistance produced by friction ...ownslope as a result of shear failure ...rm landslide is applied. This may ...ides are ...erial by ...t thereby ...the mass,

...movement ...tent than

...It then flows ...and weakened

By contrast, flows are more continuous, less jerky, and are more likely to contort the mass into a new form (Figure 3.9). The material involved is predominantly made up of fine particles, such as deeply weathered clay.

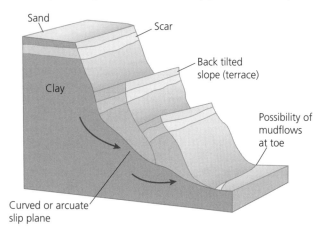

Figure 3.8 Slumps

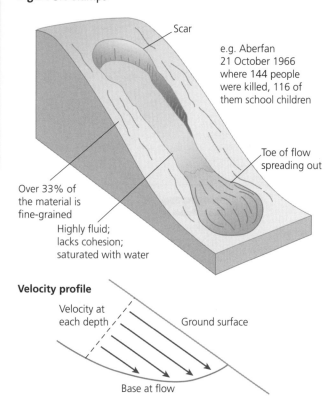

Figure 3.9 Flows

Avalanches

Avalanches are rapid movements of snow, ice, rock or earth down a slope. Snow and ice may pick up rocks and/or earth. In steep mountainous areas, a rock avalanche suggests a large-scale movement of material, whereas a rock fall could be of individual rocks.

● They are common in mountainous areas: newly fallen snow may fall off older snow, especially in winter (a dry avalanche), while in spring partially melted snow can move (a wet avalanche), often triggered by skiing (Figure 3.10).

● Avalanches frequently occur on steep slopes over 22°, especially on north-facing slopes where the lack of sun inhibits the stabilisation of snow.

● Debris avalanches are rapid mass movements of sediments, often associated with saturated ground conditions.

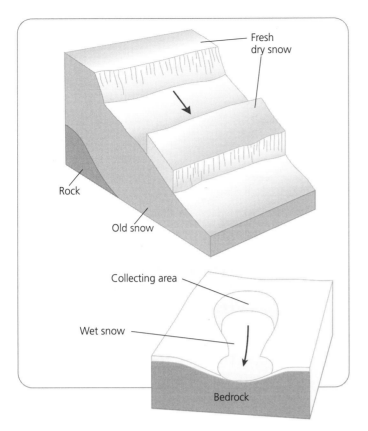

Figure 3.10 Avalanches

Now test yourself

25 Compare and contrast the characteristics of falls and slides.
26 How does a slump differ from a slide?
27 Explain the terms rotational slide and avalanche.
28 Outline how avalanches are formed.

Answers on p.215

Tested

3.4 The human impact

The human impact is of great importance. Landforms can be created by constructional activities (tipping, excavation and hydrological interference) and by farming. Hillsides have been terraced in many parts of the world for centuries.

Pollution can also affect natural processes. Pollution can be natural, such as from volcanic eruptions, as well as human in origin. It can be deliberate or accidental.

Pollution is defined as the contamination of the Earth/atmosphere system to such an extent that normal environmental processes are adversely affected. The elements involved are described as disagreeable, noxious, toxic, harmful and/or objectionable.

Weathering

Revised

Weathering processes can be intensified by changes in local climate, as shown by increased chemical weathering in urban areas. Changes in the nature and rate of weathering are closely linked to air quality.

- Increased emissions of SO_2 (from the burning of fossil fuels) have led to higher levels of sulfuric acid.
- Chemical reactions with SO_2 can create salts, such as calcium sulfate and magnesium sulfate. These are able to chemically weather rocks.

Human activity has many impacts on the nature and rate of limestone denudation:

- The burning of fossil fuels and deforestation has raised atmospheric levels of CO_2, increasing the potential for carbonation–solution.
- There are increasing levels of acidity in rain water due to SO_2 and nitrogen oxides in the atmosphere.
- Increased lighting in caves allows plants to grow, leading to biological weathering and increased levels of organic acids.

Mass movement

Revised

Rates of mass movements can be altered by building, excavation, drainage or agriculture, all of which can destabilise slopes. Some mass movements are created by humans piling up waste soil and rock into unstable accumulations that move without warning. Landslides can be created by undercutting or overloading. As well as causing mass movements, human activities can reduce them (Table 3.2).

Table 3.2 Examples of methods of controlling mass movement

Type of movement	Method of control
Falls	Grading or benching the slope to flatten it
	Drainage
	Reinforcement of rock walls by grouting with cement or using anchor bolts
	Covering of wall with steel mesh
Slides and flows	Grading or benching the slope to flatten it
	Drainage of surface water with ditches
	Sealing surface cracks to prevent infiltration
	Sub-surface drainage
	Rock or earth buttresses at foot
	Retaining walls at foot
	Pilings through the potential slide mass

Now test yourself

29 How can human activity reduce the risk of rockfalls?

Answer on p.215

Tested

In urban areas slope modification is often very significant, given the need for buildings and roads to be constructed safely. Even on flat sites, large modern buildings generally involve the removal of material to allow for proper foundations. Slope modification tends to increase as construction moves onto steeper slopes. The steep slopes, devoid of soil and vegetation, are potentially much less stable than the former natural slope and are, in times of intense rainfall, susceptible to small but quite damaging land slips.

Mining and quarrying

Revised

The environmental impacts of mining are diverse (Table 3.3):

- Habitat destruction is widespread, especially if opencast or strip mining is used. Opencast mining is a form of extensive excavation in which the overlying material (overburden) is removed by machinery, revealing the seams or deposits below.
- Disposal of waste rock and 'tailings' (the impurities left behind after a mineral has been extracted from its ore) can destroy vast expanses of ecosystems.
- Copper mining is especially polluting. Producing 1 tonne of copper creates over 100 tonnes of waste rock. Even the production of 1 tonne of china clay (kaolin) creates 1 tonne of mica, 2 tonnes of undecomposed rock and 6 tonnes of quartz sand.

Expert tip

There is widespread pollution from many forms of mining. This results from the extraction, transport and processing of the raw material, and affects air, soil and water. Water is affected by heavy metal pollution, acid mine drainage, eutrophication and deoxygenation.

Table 3.3 Environmental problems associated with mining

Problem	Type of mining operation			
	Open pit and quarrying	Opencast (e.g. coal)	Underground (e.g. tin or gold)	Dredging
Habitat destruction	✗	✗	–	✗
Dump failure/erosion	✗	✗	✗	–
Subsidence	–	–	✗	–
Water pollution	✗	✗	✗	✗
Air pollution*	✗	✗	✗	–
Noise	✗	✗	–	–
Vibration	✗	✗	–	–
Visual intrusion	✗	✗	✗	✗
Dereliction	✗	–	✗	✗

✗ Problem present
– Problem unlikely
* Can be associated with smelting, which may not be at the site of ore/mineral extraction

Now test yourself Tested

30 What are the environmental impacts of opencast mining?
31 What impacts can underground mining have?

Answers on p.215

Expert tip

The mining of coal has a number of environmental impacts. There is also a considerable impact on water resources and quality. For example underground mining uses 63–120 litres per tonne and a further 33 litres per tonne for surface waste disposal.

Acidification
Revised

Causes

Sulfur dioxide (SO_2) and nitrogen oxides (NOx) are emitted from power stations, industrial complexes, vehicles and urban areas. Some of these oxides fall directly to the ground as **dry deposition** (dry particles, aerosols and gases) close to the source (Figure 3.11). By contrast, the longer the SO_2 and NOx remain in the air the greater the chance they will be oxidised to form sulfuric acid (H_2SO_4) and nitric acid (HNO_3). These acids dissolve in cloud droplets to form **acid precipitation** (rain, snow, mist, hail) and reach the ground as **wet deposition**.

Dry deposition refers to materials transported in suspension in the air that settle on the ground as dry matter.

Acid precipitation is any precipitation with a pH lower than 5.0.

Wet deposition refers to materials transported in suspension in the air that eventually fall to the ground as precipitation.

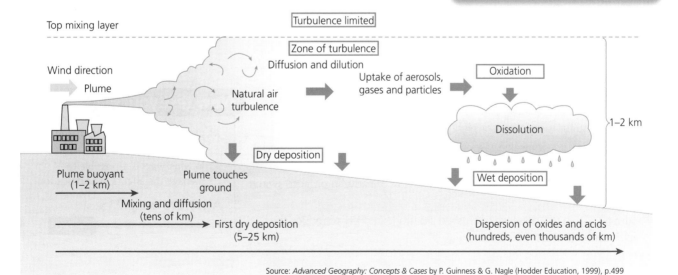

Source: *Advanced Geography: Concepts & Cases* by P. Guinness & G. Nagle (Hodder Education, 1999), p.499

Figure 3.11 Types of acid deposition

Location of impacts

The worst-hit areas used to include Scandinavia and eastern North America (Figure 3.12). The future trends are likely to see increased sulfur emissions in NICs (newly industrialising countries) such as China, India and Brazil.

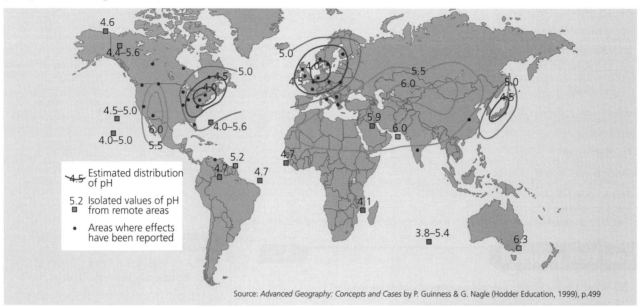

Source: *Advanced Geography: Concepts and Cases* by P. Guinness & G. Nagle (Hodder Education, 1999), p.499

Figure 3.12 Highly acidified areas worldwide

Effects of acid rain

- The first effects of acid rain were noted in Scandinavian lakes in the 1960s. Over 18,000 lakes in Sweden are acidified. Fish stocks in about 9000 Swedish lakes, mostly in the south and the centre of the country, are badly affected.
- The major health concern about of acid water is due to its ability to flush trace metals from soils into lakes and groundwater. Some wells in Sweden have aluminium levels of up to 1.7 mg/l compared with a WHO safe limit of 0.2 mg/l. High levels of metal mercury in fish can cause serious health problems when eaten by people.
- Trees and forests are severely affected by acid rain. Sulfur dioxide interferes with the process of photosynthesis. Coniferous trees seem to be most at risk from acid rain and experience needle drop.
- Acidified lakes are characterised by an impoverished species structure because harmful metals such as cadmium, copper, aluminium, zinc and lead become more soluble and are more easily available to plants and animals.
- Acid rain corrodes metal and stonework, making the maintenance of buildings more costly. The major threats are to older, historic buildings.

Possible solutions

Acid waters can be neutralised by adding lime. However, it is preferable to prevent the process in the first place, by:

- using low-sulfur fuels (oil/gas or high-grade coal)
- removing sulfur from waste gases after combustion – flue gas desulfurisation (FGD)
- burning less fossil fuel (this requires a government initiative in order to switch to nuclear or HEP)
- burning coal in the presence of crushed limestone in order to reduce the acidification process

> **Typical mistake**
>
> Some students think that acid rain is only caused by human activity – in fact, it can be entirely natural. Volcanoes emit sulfur during eruptions; for example, the acidification of Chances Peak on Montserrat is entirely due to natural causes.

> **Expert tip**
>
> Some environments are able to neutralise the affects of acid rain. Chalk and limestone areas are very alkaline and can neutralise acids very effectively. The underlying rocks over much of Scandinavia, Scotland and northern Canada are granite. They are naturally acidic, and have a very low buffering capacity.

> **Now test yourself**
>
> **32** Outline the natural and man-made causes of acidification.
>
> **33** In what ways is it possible to manage the effects of acidification?
>
> **Answers on p.215**
>
> Tested ☐

Exam-style questions

Section A

1 Define the term weathering.   [1]

2 Briefly explain how climate influences weathering.   [3]

3 Explain the formation of tors.   [6]

Section B

1 In what ways is it possible to classify mass movements?   [7]

2 Outline the factors that influence mass movements.   [8]

3 Using examples, explain how human activities can influence mass movements. [10]

Exam ready

4 Population

4.1 Natural increase as a component of population change

Population growth Revised ☐

The first hominids appeared in Africa around 5 million years ago. During most of the period of early humankind the global population was very low. 10,000 years ago it was no more than 5 million. By 3500 BC the global population had reached 30 million and by the time of Christ this had risen to about 250 million. World population reached 500 million by about 1650. From this time population grew at an increasing rate. By 1800 global population had doubled to reach 1 billion. Table 4.1 shows the time taken for each subsequent billion to be reached.

Table 4.1 World population growth by each billion

Each billion	Year	Number of years to add each billion
1st	1800	All of human history
2nd	1930	130
3rd	1960	30
4th	1974	14
5th	1987	13
6th	1999	12
7th	2011	12
8th	2024	13

Recent demographic change

- The rate of population growth is much higher in the developing world compared with the more developed world.
- However, only since the Second World War has population growth rate in the developing world overtaken that in the developed world.
- The developed countries had their period of high population growth in the nineteenth and early twentieth centuries, while for the developing countries high population growth has occurred since 1950.
- The highest ever global population growth rate was reached in the early-to-mid-1960s when population growth in the developing world peaked at 2.4% a year.
- Even though the rate of global growth has been falling for three decades, population momentum means that the numbers being added each year did not peak until the late 1980s.

The components of population change Revised ☐

Natural change is the balance between births and deaths, while net migration is the difference between immigration and emigration (Figure 4.1). The relative contributions of natural change and net migration can vary over time within a particular country. Natural change can be stated in absolute or relative terms.

The former gives the actual change in population, for example 200,000. The latter expresses natural change as a rate per thousand, for example 2/1000.

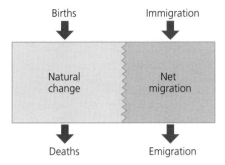

Figure 4.1 Input–output model of population change

Births Immigration

Natural change   Net migration

Deaths   Emigration

The factors affecting levels of fertility

Fertility varies widely around the world. The **crude birth rate** is the most basic measure of fertility. The word 'crude' means that the birth rate applies to the total population, taking no account of gender and age. The crude birth rate is heavily influenced by the age structure of a population. In 2010 the crude birth rate varied globally from a high of 52/1000 in Niger to a low of 7/1000 in Monaco.

For more accurate measures of fertility, the **fertility rate** and the **total fertility rate** are used. These rates apply to women in the main reproductive age range, rather than to the whole population. They are thus much more accurate measures of fertility than the birth rate. In 2010, the total fertility rate varied from a high of 7.4 in Niger to a low of 1.0 in China, Macao and China, Hong Kong. The global average is 2.5. The factors affecting fertility can be grouped into four main categories (Table 4.2).

Table 4.2 The factors affecting fertility

Demographic	Other population factors, particularly infant mortality rates, influence fertility.
Social/cultural	In some societies, particularly in Africa, tradition demands high rates of reproduction. Education, especially female literacy, is the key to lower fertility. In some countries religion is an important influence.
Economic	In many poor countries children are seen as an economic asset, while in the more developed world the general perception is reversed and the cost of the child dependency years is a major factor in the decision to begin or extend a family.
Political	There are many examples in the past century of governments attempting to change the rate of population growth for economic and strategic reasons.

Fertility decline

In 2007 the United Nations predicted that global population would peak at 9.2 billion in 2050. The global peak population has been continually revised downwards in recent decades. Fertility levels in most parts of the world have fallen faster than previously expected. In the second half of the 1960s, after a quarter century of increasing growth, the rate of world population growth began to slow down. Since then some developing countries have seen the speediest falls in fertility ever known.

A fertility rate of 2.1 children per woman is **replacement-level fertility**, below which populations eventually start falling. In 2010 there were already 87 countries with total fertility rates at or below 2.1. This number is likely to increase.

The factors affecting mortality

The **crude death rate** is heavily influenced by the age structure of a population. For example, the crude death rate for the UK is 9/1000 compared with 6/1000 in Brazil. Yet life expectancy at birth in the UK is 80 years compared with 73 years

---

Crude birth rate is the number of live births per 1000 population in a given year.

Natural change is the difference between the number of births and the number of deaths in a country or region. This can be positive (natural increase) or negative (natural decrease).

Typical mistake

The crude birth rate is often presented as an accurate measure of fertility. However, it is only a very broad indicator as it does not take into account the age and sex distribution of the population.

Fertility rate is the number of live births per 1000 women aged 15–49 years in a given year.

Total fertility rate is the average number of children that would be born alive to a woman (or group of women) during her lifetime, if she were to pass through her child-bearing years conforming to the age–specific fertility rates of a given year.

Replacement-level fertility is the level at which those in each generation have just enough children to replace themselves in the population. A total fertility rate of 2.1 children is usually considered as replacement level.

Now test yourself

1. Define (a) the crude birth rate and (b) the crude death rate.
2. How is the rate of natural increase calculated?
3. Why is the fertility rate a better measure of fertility than the crude birth rate?
4. List the four general factors affecting fertility.

Answers on p.215

Tested

Crude death rate is the number of deaths per 1000 population in a given year.

Infant mortality rate is the number of deaths of infants under 1 year of age per 1000 live births in a given year.

Life expectancy (at birth) is the average number of years a person may expect to live when born, assuming past trends continue.

in Brazil. In 2010 the crude death rate varied around the world from a high of 20/1000 in Zambia to a low of 1/1000 in Qatar. The **infant mortality rate** and **life expectancy** are much more accurate measures of mortality.

The causes of death vary significantly between the more developed and less developed worlds:

● In the developing world, infectious and parasitic diseases account for over 40% of all deaths.

● When people live in overcrowded and insanitary conditions communicable diseases such as tuberculosis and cholera can spread rapidly.

● Limited access to health care and medicines means that otherwise treatable conditions such as malaria and tuberculosis are often fatal to poor people.

● Poor nutrition and deficient immune systems are also key risk factors for several big killers such as lower respiratory infections, tuberculosis and measles.

● In rich countries heart disease and cancer are the big killers.

Infant mortality

There is a huge contrast in infant mortality by world region. In 2010 Africa had the highest rate (76/1000) and Europe and North America the lowest rate (6/1000). The infant mortality rate is frequently considered to be the most sensitive indicator of socio-economic progress, being heavily influenced by fundamental improvements in quality of life factors such as water supply, nutrition and health care.

Life expectancy

● In 2010 the lowest average life expectancy by world region was in Africa (55 years), with the highest in North America (78 years).

● The range in life expectancy among rich and poor countries has narrowed significantly during the last 50 years or so.

● However, the impact of AIDS in particular has caused recent decreases in life expectancy in some countries in sub-Saharan Africa.

The interpretation of age/sex pyramids

Revised

The most studied aspects of **population structure** are age and sex. Other aspects of population structure that can be studied include race, language, religion and social/occupational groups.

The age and sex structure is illustrated by the use of a **population pyramid**, which can show either absolute or relative data. Figure 4.2 provides some useful tips for understanding population pyramids.

> **Population structure** is the composition of a population, the most important elements of which are age and sex.
>
> A **population pyramid** is a bar chart, arranged vertically, that shows the distribution of a population by age and sex.

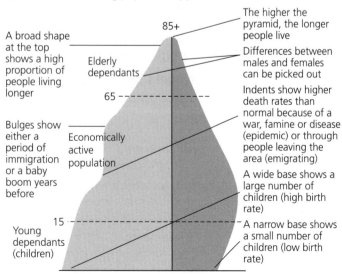

Figure 4.2 An annotated population pyramid

Population pyramids change significantly in shape as a country progresses through demographic transition (Figure 4.3).

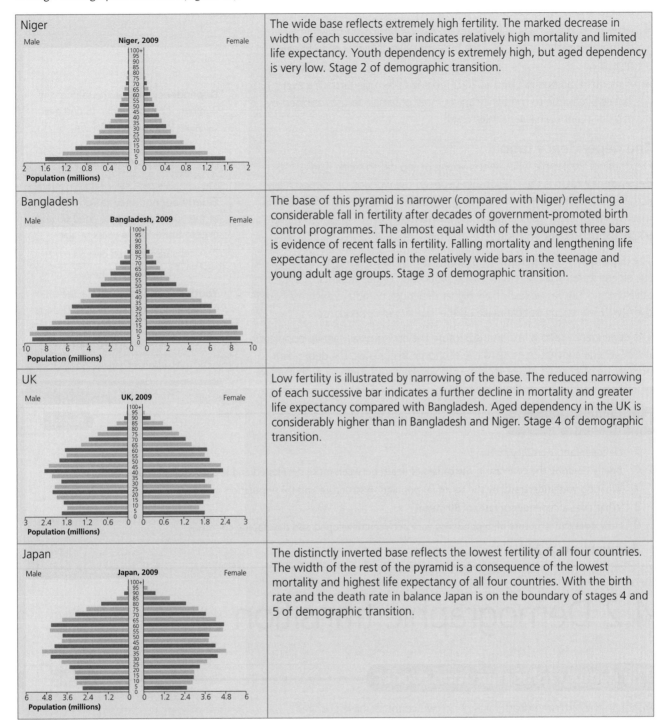

Niger	The wide base reflects extremely high fertility. The marked decrease in width of each successive bar indicates relatively high mortality and limited life expectancy. Youth dependency is extremely high, but aged dependency is very low. Stage 2 of demographic transition.
Bangladesh	The base of this pyramid is narrower (compared with Niger) reflecting a considerable fall in fertility after decades of government-promoted birth control programmes. The almost equal width of the youngest three bars is evidence of recent falls in fertility. Falling mortality and lengthening life expectancy are reflected in the relatively wide bars in the teenage and young adult age groups. Stage 3 of demographic transition.
UK	Low fertility is illustrated by narrowing of the base. The reduced narrowing of each successive bar indicates a further decline in mortality and greater life expectancy compared with Bangladesh. Aged dependency in the UK is considerably higher than in Bangladesh and Niger. Stage 4 of demographic transition.
Japan	The distinctly inverted base reflects the lowest fertility of all four countries. The width of the rest of the pyramid is a consequence of the lowest mortality and highest life expectancy of all four countries. With the birth rate and the death rate in balance Japan is on the boundary of stages 4 and 5 of demographic transition.

Figure 4.3 Four population pyramids

In countries where there is strong rural-to-urban migration, the population structures of the areas affected can be markedly different. These differences show up clearly on population pyramids. Out-migration from rural areas is age-selective, with single young adults and young adults with children dominating this process. Thus, the bars for these age groups in rural areas affected by out-migration will indicate fewer people than expected. In contrast, the population pyramids for urban areas attracting migrants will show age-selective in-migration, with substantially more people in these age groups than expected. Such migrations may also be sex-selective.

Expert tip

When analysing population pyramids a good starting point is to divide the pyramid into three sections: the young dependent population, the economically active population, and the elderly dependent population. You can then comment on each section in turn.

The **sex ratio** is the number of males per 100 females in a population:

- Male births consistently exceed female births due to a combination of biological and social reasons.
- After birth, the gap generally begins to narrow until eventually females outnumber males, as at every age male mortality is higher than female mortality.
- A report published in China in 2002 recorded 116 male births for every 100 female births due to the significant number of female fetuses aborted by parents intent on having a male child.

The dependency ratio

Dependants are people who are too young or too old to work. The **dependency ratio** is the relationship between the working or economically active population and the non-working population. A dependency ratio of 60 means that for every 100 people in the economically active population there are 60 people dependent on them.

The dependency ratio in developed countries is usually between 50 and 75 with the elderly forming an increasingly high proportion of dependants. In contrast, developing countries typically have higher dependency ratios, which may reach over 100. Here young people make up the majority of dependants.

The dependency ratio is important because the economically active population will in general contribute more to the economy. In contrast, the dependent population tend to be bigger recipients of government funding, particularly for education, health care and public pensions.

> **Dependency ratio** is the ratio of the number of people under 15 and over 64 years to those aged 15–64.
>
> $$\frac{\text{number aged 0–14} + \text{number aged over 64}}{\text{number aged 15–64}} \times 100$$
>
> **Elderly dependency ratio** is the ratio of the number of people aged 65 and over to those 15–64 years of age.
>
> $$\frac{\text{number aged over 64}}{\text{number aged 15–64)}} \times 100$$
>
> **Youth dependency ratio** is the ratio of the number of people aged 0–14 to those 15–64 years of age.
>
> $$\frac{\text{number aged 0–14}}{\text{number aged 15–64}} \times 100$$

Now test yourself

Tested

5 Define infant mortality rate.

6 Briefly describe the contrast in the causes of death between more developed and less developed countries.

7 What do you understand by the terms (a) population structure and (b) population pyramid?

8 What does a dependency ratio of 80 mean?

9 How does the structure of dependency vary between developed and developing countries?

Answers on p.215

4.2 Demographic transition

The demographic transition model

Revised

Although the birth and death rates of no two countries have changed in exactly the same way, some broad generalisations can be made about population change that are illustrated by the model of **demographic transition** (Figure 4.4).

> **Demographic transition** is the historical shift of birth and death rates from high to low levels in a population.

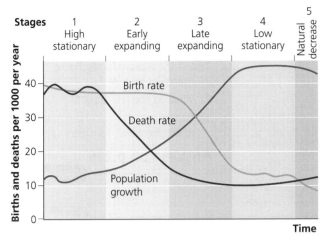

Figure 4.4 Model of demographic transition

- No country as a whole retains the characteristics of stage 1, which applies only to the most remote societies on Earth.
- The poorest of the developing countries are in stage 2.
- Most developing countries that have registered significant social and economic advances are in stage 3.
- All the developed countries of the world are now in stages 4 or 5.

The basic characteristics of each stage are as follows:

Stage 1 The crude birth rate is high and stable while the crude death rate is high and fluctuating due to the sporadic incidence of famine, disease and war. Population growth is very slow and there may be periods of considerable decline. Infant mortality is high and life expectancy low.

Stage 2 The death rate declines significantly. The birth rate remains high as the social norms governing fertility take time to change. The rate of natural change increases to a peak at the end of this stage. Infant mortality falls and life expectancy increases. The reasons for the decline in mortality are: better nutrition; improved public health, particularly in terms of clean water supply and efficient sewerage systems; and medical advances.

Stage 3 Social norms adjust to the lower level of mortality and the birth rate begins to decline. Life expectancy continues to increase and infant mortality to decrease.

Stage 4 Both birth and death rates are low. The former is generally slightly higher, fluctuating somewhat due to changing economic conditions. Population growth is slow. Death rates rise slightly as the average age of the population increases. However, life expectancy still improves as age-specific mortality rates continue to fall.

Stage 5 The birth rate has fallen below the death rate, resulting in **natural decrease**. In the absence of net migration inflows these populations are declining.

Criticisms of the model:

- It is seen as too Eurocentric as it was based on the experience of Western Europe.
- Many developing countries may not follow the sequence set out in the model.
- It fails to take into account changes due to migration.

Demographic transition in the developing world

Compared with the experiences of most developed nations before them, in the developing world:

- birth rates in stages 1 and 2 were generally higher
- the death rate fell much more steeply and for different reasons
- some countries had much larger base populations and thus the impact of high growth in stage 2 and the early part of stage 3 has been far greater
- for those countries in stage 3 the fall in fertility has also been steeper

Issues of ageing populations
Revised

- The world's population is ageing significantly.
- The global average for life expectancy increased from 46 years in 1950 to nearly 65 in 2000. It is projected to reach 74 years by 2050.
- Europe is the 'oldest' region in the world. Those aged 60 years and over currently form 20% of the population.
- Africa is the 'youngest' region in the world, with the proportion of children accounting for 43% of the population today.

The problem of demographic ageing has been a concern of developed countries for some time, but it is now also beginning to alarm developing nations. Demographic ageing will put health-care systems, public pensions, and government budgets in general, under increasing pressure. In many developed countries the fastest-growing segment of the population is the so-called 'oldest-old' (85 years or more). It is this age group that is most likely to need expensive residential care.

Japan has the most rapidly ageing population in history. It has the highest life expectancy in the world and one in five of the population is over the age of 60. No other country has a lower percentage of its population under 15. Younger workers are at a premium and there is considerable competition to recruit them.

> An **ageing population** is one undergoing a rise in its median age. It occurs when fertility declines, while life expectancy remains constant or increases.

The link between population and development
Revised

Development involves improvement in the quality of life. This includes wealth, but it also refers to other important aspects of our lives, such as education, health and freedom of expression. For example, development occurs in a low-income country when:

- local food supply improves due to investment in machinery and fertilisers
- the electricity grid extends outwards from the main urban areas to rural areas

> **Development** is the use of resources to improve the quality of life in a country.

The human development index (HDI)

The highest-ranking countries according to the **human development index** are in stage 4 (or stage 5) of demographic transition, suggesting a very strong link between the rate of population growth and the level of economic development. The countries with low human development invariably have high rates of population growth and most are in stage 2 of demographic transition. The more advanced developing countries are generally in stage 3 of demographic transition. This includes countries such as Brazil, Mexico, India and Malaysia. However, the development process is complex and is the result of the interaction of a wide range of factors.

Changes in demographic indices over time

Figure 4.5 illustrates changes in birth and death rates in England and Wales between 1700 and 2000. The birth and death rates in stages 1, 2 and 3 broadly

> **Expert tip**
>
> You should be able to differentiate between economic growth and development. The former is an increase in GDP while the latter is a much more wide ranging concept concerning many more aspects of quality of life, such as education and health.

> The **human development index** is a measure of development that combines three important aspects of human well-being: life expectancy, education and income.

correspond to those in many poorer societies today. The graph identifies a range of important factors that influenced birth and death rates in England and Wales during the time period concerned.

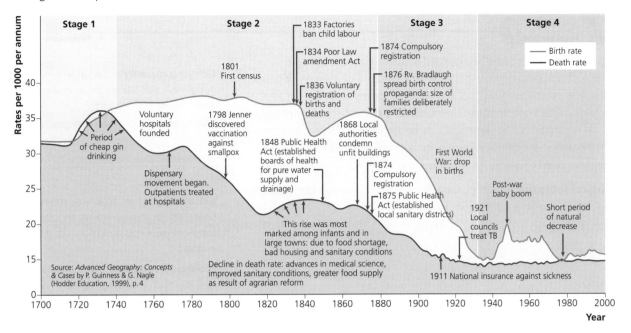

Figure 4.5 England and Wales: changes in birth and death rates, 1700–2000

Child mortality has been falling significantly. Globally, the number of children under 5 years who died in 2006 dropped below 10 million for the first time, down from almost 13 million in 1990. The main reasons for the decline included measles vaccinations, mosquito nets and increased rates of breast-feeding. Unicef argues that the majority of the remaining child deaths are preventable.

Reducing **maternal mortality** is one of the UN's eight Millennium Development Goals. There is a huge contrast in maternal mortality between the developed and developing worlds. Globally, 1 in 92 women die from pregnancy-related causes. However, in more developed nations the risk is only 1 in 6000 compared with 1 in 22 in sub-Saharan Africa. Major influencing factors in maternal mortality are the type of pre-natal care and the type of attendance at birth.

On a global scale 75% of the total improvement in longevity has been achieved in the twentieth century and the early years of the twenty-first century. In 1900 the world average for life expectancy is estimated to have been about 30 years, but by 1950–55 it had risen to 46 years. The current global average is 68 years.

> **Typical mistake**
>
> Students sometimes confuse child mortality and infant mortality. The former refers to children under the age of 5, the latter to children under the age of 1.

> **Child mortality rate** is the number of children who die before their 5th birthday per 1000 live births.
>
> **Maternal mortality** is the death of a woman during or shortly after a pregnancy.

> **Now test yourself**   Tested
>
> **14** By how much has global average life expectancy increased since 1950?
> **15** Which **two** factors have resulted in such an elderly population in Japan?
> **16** Which three quality-of-life indicators are used in the human development index?
> **17** Define child mortality.
> **18** By how much did child mortality decrease between 1990 and 2006?
>
> **Answers on p.215**

4.3 Population–resource relationships

Carrying capacity

Revised

Carrying capacity is a dynamic, as opposed to static, concept because advances in technology can increase the carrying capacity of a region or country significantly. The enormous growth of the global economy in recent decades has had a huge impact on the planet's resources and natural environment. Many resources are running out and waste sinks are becoming full. Climate change will impact on a number of essential resources for human survival, increasing the competition between countries for such resources.

> **Carrying capacity** is the largest population that the resources of a given environment can support.

The **ecological footprint** is an important measure of humanity's demands on the natural environment. It has six components:

- built-up land
- fishing ground
- forest
- grazing land
- cropland
- carbon footprint

An ecological footprint is measured in **global hectares**. Nations at different income levels show considerable disparities in the extent of their ecological footprint. In 1961, most countries in the world had more than enough **biocapacity** to meet their own demand. But by the mid-1980s humankind's ecological footprint had reached the Earth's biocapacity. Since then humanity has been in ecological 'overshoot'.

> **Biocapacity** is the capacity of an area or ecosystem to generate an ongoing supply of resources and to absorb its wastes.
>
> **Ecological footprint** is a sustainability indicator that expresses the relationship between population and the natural environment. It takes into account the use of natural resources by a country's population.
>
> One **global hectare** is equivalent to one hectare of biologically productive space with world average productivity.

In many countries the **carbon footprint** is the dominant element of the six components that comprise the ecological footprint, while in some, other aspects of the ecological footprint are more important. In general the relative importance of the carbon footprint declines as the total ecological footprint of countries falls.

- The ecological footprint is strongly influenced by the size of a country's population.
- The other main influences are the level of demand for goods and services in a country (the standard of living), and how this demand is met in terms of environmental impact.
- International trade is taken into account in the calculation of a country's ecological footprint. For each country its imports are added to its production while its exports are subtracted from its total. The expansion of world trade has been an important factor in the growth of humanity's total ecological footprint.

> **Typical mistake**
>
> Sometimes students think that the ecological footprint and the carbon footprint are the same. However, the carbon footprint is only one component of the ecological footprint, even though for many countries it is the most important component.

> **Carbon footprint** is defined as 'the total set of GHG (greenhouse gas) emissions caused directly and indirectly by an individual, organisation, event or product' (UK Carbon Trust 2008).

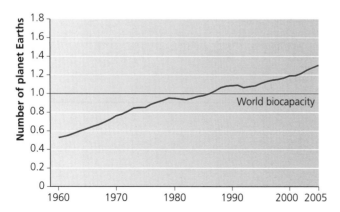

Figure 4.6 Global ecological footprint, 1960–2005

Now test yourself

19 Define carrying capacity.
20 What is the ecological footprint?
21 List the six components of the ecological footprint.
22 Define biocapacity.
23 Describe the trend illustrated in Figure 4.6.

Answers on pp.215–216

Tested

Figure 4.6 shows how humanity's ecological footprint increased from 1960 to 2005. The global ecological footprint now exceeds the planet's regenerative capacity by about 30%. This global excess is increasing and as a result ecosystems are being run down and waste is accumulating in the air, land and water. The resulting deforestation, water shortages, declining biodiversity and climate change are putting the future development of all countries at risk.

The causes and consequences of food shortages

Revised

About 800 million people in the world suffer from hunger. The problem is mainly concentrated in Africa. Food shortages can occur because of both natural and human problems. The natural problems that can lead to food shortages include:

- soil exhaustion
- drought
- floods
- tropical cyclones
- pests
- disease

However, economic and political factors can also contribute to food shortages. Such factors include:

- low capital investment
- rapidly rising population
- poor distribution/transport difficulties
- conflict situations

The effects of food shortages are both short and long term. Malnutrition can affect a considerable number of people, particularly children, within a relatively short period when food supplies are significantly reduced. With malnutrition people are more prone to disease and likely to fall ill. Such diseases include beri-beri (vitamin B1 deficiency) and kwashiorkor (protein deficiency). People who are continually starved of nutrients never fulfil their physical or intellectual potential. Malnutrition reduces people's capacity to work. This is threatening to lock parts of the developing world into an endless cycle of ill health, low productivity and underdevelopment.

The role of technology and innovation in resource development

Revised

The global usage of resources has changed dramatically over time. Technological advance has been the key to the development of new resources and the replacement of less efficient with more efficient resources. Examples of technological development in the UK include:

- the development of the nuclear power industry, which found a new use for uranium, significantly increasing its price
- renewable energy technology, particularly the construction of offshore wind farms, which is now beginning to utilise flow resources in a significant way

Innovation in food production has been essential to feeding a rising global population. The package of agricultural improvements generally known as the **Green Revolution** was seen as the answer to the food problem in many parts of the developing world. India was one of the first countries to benefit when a high-yielding variety seed programme (HVP) commenced in 1966–67. The HVP introduced new hybrid varieties of five cereals: wheat, rice, maize, sorghum and millet.

The **Green Revolution** refers to the introduction of high-yielding seeds and modern agricultural techniques in developing countries.

Table 4.3 The advantages and disadvantages of the Green Revolution

Advantages	Disadvantages
Yields are twice to four times greater than traditional varieties	High inputs of fertiliser and pesticide are required to optimise production; this is costly in both economic and environmental terms
The shorter growing season has allowed the introduction of an extra crop in some areas	HYVs require more weed control and are often more susceptible to pests and disease
Farming incomes have increased, allowing the purchase of machinery, better seeds, fertilisers and pesticides	Middle- and higher-income farmers have often benefited much more than the majority on low incomes
The diet of rural communities is now more varied	Mechanisation has increased rural unemployment
Local infrastructure has been upgraded to accommodate a stronger market-based approach	The problem of salinisation has increased with the expansion of irrigation
Employment has been created in industries supplying farms with inputs	Some HYVs have an inferior taste
Higher returns have justified a significant increase in irrigation	HYVs can be low in minerals and vitamins

The role of constraints in sustaining populations

Revised

There are a significant number of potential constraints in developing resources to sustain changing populations. Figure 4.7 illustrates the factors affecting the development of a particular resource body. The factors included in the diagram are those that operate in normal economic conditions.

Figure 4.7 Factors affecting the development of a particular resource body

War is a major issue for development. It significantly retards development and the ability of a country to sustain its population. Major conflict can set back the process of development by decades. Trade barriers form another significant constraint. Many developing countries complain that the trade barriers erected by many developed countries are too high. This reduces the export potential of poorer countries and hinders development. Climatic and other hazards in the short term and climate change in the medium and long term impact seriously on the utilisation of resources.

Overpopulation, optimum population and underpopulation

The idea of **optimum population** has been mainly understood in an economic sense (Figure 4.8). At first, an increasing population allows for a fuller exploitation of a country's resource base, causing living standards to rise. However, beyond a certain level rising numbers place increasing pressure on resources and living standards begin to decline. The highest average living standard marks the optimum population or, more accurately the **economic optimum**. Before that population is reached, the country or region can be said to be **underpopulated**. As the population rises beyond the optimum, the country or region can be said to be **overpopulated**.

> **Optimum population** is one that achieves a given aim in the most satisfactory way.
>
> **Economic optimum** is the level of population that, through the production of goods and services, provides the highest average standard of living.
>
> **Underpopulated** – when there are too few people in an area to use the resources available efficiently.
>
> **Overpopulated** – when there are too many people in an area relative to the resources and the level of technology available.

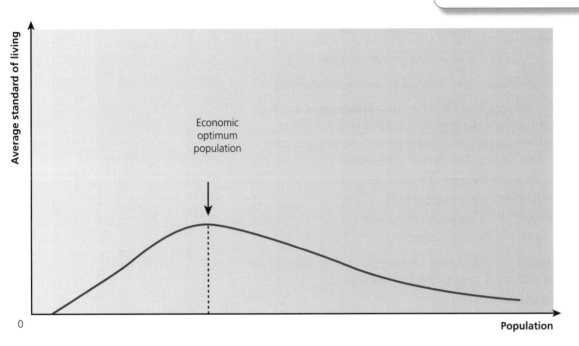

Figure 4.8 The optimum population

The most obvious examples of **population pressure** are in the developing world but the question here is: are these cases of absolute overpopulation or the results of underdevelopment that can be rectified by adopting remedial strategies over time?

The ideas of Thomas Malthus

- The Rev Malthus produced his *Essay on the Principle of Population* in 1798.
- He maintained that while the supply of food could, at best, only be increased by a constant amount in arithmetical progression (1–2–3–4–5–6…), the human population tends to increase in geometrical progression (1–2–4–8–16–32…).
- In time, population would outstrip food supply until a catastrophe occurred in the form of famine, disease or war.
- These limiting factors maintained a balance between population and resources in the long term.

Malthus could not have foreseen the great advances that were to unfold in the following two centuries. However, nearly all of the world's productive land is already exploited. Most of the unexploited land is either too steep, too wet, too dry or too cold for agriculture.

There are two opposing views of the effects of population growth:

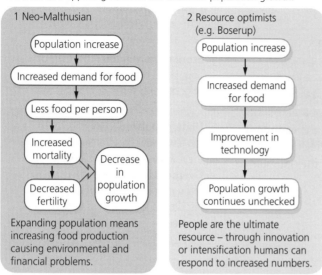

Source: *Advanced Geography: Concepts & Cases*
by P. Guinness & G. Nagle (Hodder Education, 1999), p.35

Figure 4.9 The opposing views of the neo-Malthusians and the anti-Malthusians

Figure 4.9 summarises the opposing views of the **neo-Malthusians** and the resource optimists such as Esther Boserup. Neo-Mathusians argue that an expanding population will lead to unsustainable pressure on food and other resources. In recent years neo-Malthusians have highlighted:

● the steady global decline in the area of farmland per person
● the steep rise in the cost of many food products
● the growing scarcity of fish in many parts of the world
● the continuing increase in the world's population

The **anti-Malthusians** or resource optimists believe that human ingenuity will continue to conquer resource problems. They have highlighted:

● the development of new resources
● the replacement of less efficient with more efficient resources
● the rapid development of green technology, with increasing research and development in this growing economic sector

The concept of a population ceiling and population adjustments over time

Studies of the growth of animal and fungus populations show that population numbers may either crash after reaching a high level or reach an equilibrium around the carrying capacity. These contrasting scenarios are represented by S and J growth curves. Both incorporate the concept of a population ceiling beyond which a population cannot grow because of the influence of limiting factors such as lack of food, limited space and disease.

S-curves begin with exponential growth, but beyond a certain population size the growth rate gradually slows, eventually resulting in a stable population. J-curves illustrate a 'high growth and collapse' pattern:

● The population initially grows exponentially.
● Then the population suddenly collapses. Such collapses are known as 'diebacks'. Often the population exceeds the carrying capacity (overshoot) before the collapse occurs.

Now test yourself

24 List **four** natural problems that can lead to food shortages.
25 State **two** advantages and **two** disadvantages of the Green Revolution.
26 Define optimum population.
27 Show in numerical form how arithmetical progression differs from geometrical progression.
28 State the three limiting factors identified by Malthus.

Answers on p.216

Tested

4.4 The management of natural increase

Population policy encompasses all of the measures taken by a government aimed at influencing population size, growth, distribution or composition. Such policies may promote large families (**pro-natalist policies**) or immigration to increase its size, or encourage limitation of births (**anti-natalist policies**) to decrease it.

> **Population policy** is a government's stated aim on an aspect of its population, and the measures undertaken to achieve that aim.
>
> **Pro-natalist policy** is a population policy that aims to encourage more births through the use of incentives.
>
> **Anti-natalist policy** is a population policy designed to limit fertility through the use of both incentives and deterrents.

Case study | **Managing natural increase in China**

China, with a population in excess of 1.3 billion, has been operating the world's most strict **family planning programme**, the 'one-child policy', since 1979. Some organisations, including the UN Fund for Population Activities, have praised China's policy on birth control. Many others see it as a fundamental violation of **civil liberties**.

China's policy is based on a reward and penalty approach. Rural households that obey family planning rules get priority for loans, materials, technical assistance and social welfare. The slogan in China is, 'shao sheng kuai fu' – 'fewer births, quickly richer'. The one-child policy has been most effective in urban areas where the traditional bias of couples wanting a son has been significantly eroded. However, the story is different in rural areas where the strong desire for a male heir remains the norm. In most provincial rural areas, government policy has now relaxed so that couples can now have two children without penalties.

Chinese demographers say that the one-child policy has been successful in preventing at least 300 million births, and has played a significant role in the country's economic growth.

Between 1950 and 2005 the crude birth rate fell from 43.8/1000 to 13.6/1000 (Figure 4.10). China's birth rate is now at the level of many developed countries such as the UK.

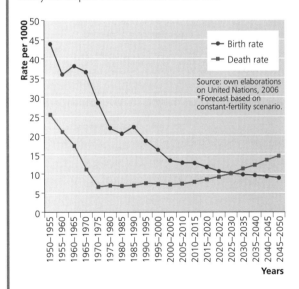

Source: own elaborations on United Nations, 2006
*Forecast based on constant-fertility scenario.

Figure 4.10 Graph of birth and death rates in China, 1950–2050

The one-child policy has brought about a number of adverse consequences including:
- demographic ageing
- an unbalanced sex ratio
- a generation of 'spoiled' children with no siblings
- a social divide, as an increasing number of wealthy couples 'buy their way round' the legislation

The policy has had a considerable impact on the sex ratio, which at birth in China is currently 119 boys to 100 girls. This compares with the natural rate of 106:100. **Selective abortion** after pre-natal screening is a major cause of the wide gap between the actual rate and the natural rate. In recent years, reference has been made to the 'four-two-one' problem whereby one adult child is left with having to provide support for his or her two parents and four grandparents. Care for the elderly is clearly going to become a major problem for the Chinese authorities, since the only social security system for most of the country's poor is their family.

While the one-child policy was introduced in 1979, this was not the first time China had tried to reduce fertility. The country's first birth control programme was introduced in 1956.

With case studies, try to make useful comparisons where appropriate. An example in this case study is that China's birth rate in 2005 was 13.6/1000 – a rate similar to that of the UK.

Now test yourself

Tested

29 Define the term population policy.

30 State **two** reasons why some countries are concerned that their fertility is too low.

31 When did China introduce the one-child policy?

32 How did China's birth rate change between 1950 and 2005?

33 State **three** adverse consequences of the one-child policy.

Answers on p.216

A **family planning programme** regulates the number and spacing of children in a family through the practice of contraception or other methods of birth control.

Civil liberties are the rights and freedoms that protect an individual from the state. Civil liberties set limits on government so that its members cannot abuse their power and interfere unduly with the lives of private citizens.

Selective abortion is an abortion performed because of the gender of the fetus or when a genetic test is performed that detects an undesirable trait.

Exam-style questions

Section A

Country	Crude birth rate (per 1000)	Crude death rate (per 1000)	Rate of natural increase (%)
USA	14	8	0.6
Japan	?	9	0.0
India	23	?	1.5
China	12	7	?
South Africa	?	12	0.9

Source: selected data from the 2011 World Population Data Sheet

1 Insert on the table the four items of missing data. [2]

2 Suggest **three** reasons for the considerable variations in birth rate between the five countries [3]

3 With reference to examples, examine the ways in which the age/sex pyramid of a typical MEDC would be different from that of a typical LEDC. [5]

Section C

Study Figure 4.4, which illustrates the model of demographic transition.

1 Describe and explain the changes that occur in stage 2. [7]

2 **(a)** What are the characteristics of stage 5?

  (b) Identify **one** country in stage 5 and suggest reasons for its demographic characteristics. [8]

3 Examine the differences between demographic transition in the MEDCs in the nineteenth and early twentieth centuries and trends in LEDCs today. [10]

Exam ready

5 Migration

5.1 Migration as a component of population change

Revised

Migration: basic terminology

The terms **immigration** and **emigration** are used with reference to international migration. The corresponding terms for internal movements are **in-migration** and **out-migration**. **Net migration** is the number of migrants entering a region or country minus the number of migrants who leave the same region or country.

Migrations are embarked upon from an area of **origin** and are completed at an area of **destination**. As E. S. Lee's model shows (Figure 5.1), there are intervening obstacles and opportunities between points of origin and destination. Migrants sharing a common origin and destination form a **migration stream**. For every migration stream a **counterstream** or reverse flow usually occurs. Push and pull factors (Figure 5.2) encourage people to migrate. **Push factors** are the negative factors at the area of origin; **pull factors** are the positive factors at the destination.

A basic distinction is between voluntary and forced migration. **Voluntary migration** is where the individual or household has a free choice about whether to move or not. **Forced migration** occurs when the individual or household has little or no choice but to move. This may be due to environmental or human factors.

> **Migration** is the movement of people across a specified boundary, national or international, to establish a new permanent place of residence. The United Nations defines permanent as a change of residence lasting more than 1 year.

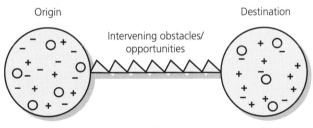

Origin   Intervening obstacles/ opportunities   Destination

+ Positive factors   − Negative factors

O Factors perceived as unimportant to the individual

Source: *IGCSE Geography* by P. Guinness & G. Nagle (Hodder Education, 2009), p.23

Figure 5.1 E.S.Lee's migration model

Typical mistake

Students sometimes confuse immigration and emigration (the terms used for crossing international borders) with in-migration and out-migration (internal movements within one country).

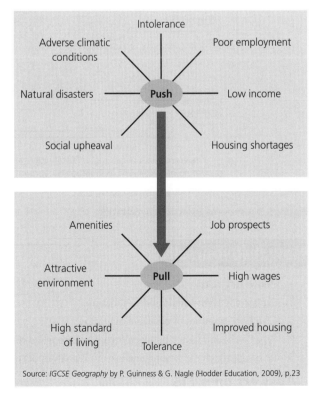

Source: *IGCSE Geography* by P. Guinness & G. Nagle (Hodder Education, 2009), p.23

Figure 5.2 Push and pull factors

Causes of migration

In 1958 W. Peterson noted the following five migratory types:

- **Primitive migration**: nomadic pastoralism and shifting cultivation practised by the world's most traditional societies.
- **Forced migration**: the abduction and transport of Africans to the Americas as slaves was the largest forced migration in history. Migrations may also be forced by natural disasters.
- **Impelled migration**: takes place under perceived threat, either human or physical, but an element of choice (lacking in forced migration) remains.
- **Free and mass migration**: both involve freedom of choice. The distinction is one of magnitude only. The movement of Europeans to North America was the largest mass migration in history.

Expert tip

Classifications such as the Peterson classification of migration should be an essential part of your core knowledge because they are important in offering logical explanations. Make sure you understand each element of the classification and that you can provide good examples of each.

Akin Mabogunje, in his analysis of rural–urban migration in Africa, used a systems approach showing migration as a circular, interdependent and self-modifying system (Figure 5.3). The system and the environment act and react upon each other continuously. For example, expansion in the urban economy will stimulate migration from rural areas while deteriorating economic conditions in the larger urban areas will result in a reduction of migration flows from rural areas. The flow of information between out-migrants and their rural origin is an important component of the system.

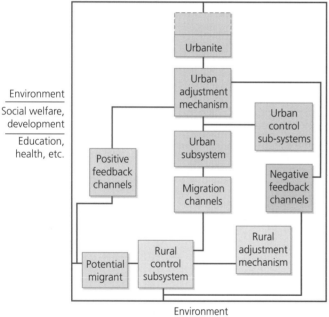

Environment
Economic conditions — wages, prices, consumer preferences, degrees of commercialisation and industrial development

Environment
Social welfare, development
Education, health, etc.

Environment
Technology
Transportation, communication, mechanisation, etc.

Environment
Governmental policies, agricultural practices, marketing organisations, population movement, etc.

Source: *Access to Geography: Migration* by P. Guinness (Hodder Education, 2002), p.18

Figure 5.3 A systems approach to migration

Now test yourself

1 Define migration.
2 Explain the terms origin and destination.
3 Give **two** examples of primitive migration.
4 What is the difference between forced and impelled migration?
5 What approach did Mabogunje use in his analysis of rural–urban migration in Africa?

Answers on p.216

Recent approaches to migration

The Todaro model: the cost–benefit approach

The simplistic explanation put forward for rural–urban migration was that many rural dwellers had been attracted by the 'bright lights' of the large urban areas without any clear understanding of the real deprivation of urban life. Michael Todaro challenged this view arguing that migrants' perceptions of urban life were realistic, being strongly based on an accurate flow of information from earlier migrants. They were very aware that in the short-term they might not be better off, but weighing up the odds the likelihood was that their socio-economic standing would improve in the long term.

Stark's 'new economics of migration'

Stark extended the Todaro model by replacing the individual with the household as the unit of analysis. Migration, according to Stark, is seen as a form of economic diversification by families whereby the costs and rewards are shared. It is a form of risk spreading.

Marxist/structuralist theory

Some writers, often in the tradition of Marxist analyses, see labour migration as inevitable in the transition to capitalism. Migration is the only option for survival after alienation from the land. Structuralist theory draws attention to the advantages of migrant labour for capitalist production and emphasises the control that capitalism has over migrant labour.

Structuration theory

Structuration theory incorporates both individual motives for migration and the structural factors in which the migrants operate. It stresses that rules designed to regulate behaviour also provide opportunity and room for manoeuvre for those they seek to constrain. This approach also builds in an awareness of cultural factors.

Gender analyses

In recent decades gender has come to occupy a significant place in migration literature. There is now much more emphasis on the different migration responses by men and women and on gender discrimination.

Case study **Push and pull factors in Brazil**

Significant rural-to-urban migration has occurred in Brazil since the 1950s, with large migration streams to the big cities such as São Paulo and Rio de Janiero (Table 5.1).

Table 5.1 Push and pull factors in Brazil

Push factors	Pull factors
The mechanisation of agriculture, which has reduced the demand for farm labour	A greater likelihood of paid employment, even if only in the informal sector
The amalgamation of farms and estates, particularly by agricultural production companies, resulting in a high level of rural–urban migration	Greater proximity to health and education services; this factor is particularly important for migrants with children
The generally poor conditions of rural employment	Better housing opportunities; even favela housing may be better than that found in some rural areas
Desertification in the northeast and deforestation in the north	Greater access to retail services than in rural areas
Unemployment and underemployment	The cultural and social attractions of large cities
Poor social conditions, particularly in terms of housing, health and education	Access to internet services, often lacking in rural areas; this is often an important factor for younger migrants

The role of constraints, obstacles and barriers

Revised

In most countries there are no legal restrictions on internal migration. Thus the main constraints are distance and cost. In contrast, immigration laws present the major barrier in international migration.

The economic costs of migration can be viewed in three parts:
- 'closing up' at the point of origin
- the actual cost of movement itself
- the costs of 'opening up' at the point of destination

The consideration of 'distance' usually involves the dangers associated with the journey. Physical factors include risks such as flood, drought, landslide and crossing water bodies. Human factors centre around any hostility from other people that may be encountered on the journey and the chances of an accident while travelling.

In terms of international migration, government attitudes in the form of immigration laws usually present the most formidable barrier to prospective migrants. Over time the legal barriers to immigration have generally become more formidable. Most countries favour immigration applications from people with skills that are in short supply and from people who intend to set up businesses and create employment.

Migration data

Revised

There are three principal sources of migration data:
- **Population censuses** are important sources of information because they are taken at regular intervals and cover whole countries. The two sorts of data generally provided are (1) birthplaces of the population and (2) period migration figures (movement over a particular period of time).
- Japan and a number of European countries (including Norway, Sweden and Switzerland) collect 'continuous data' on migration through **population registers**. Inhabitants are required to register an address with the police or a civic authority and to notify all changes of residence.
- **Specific surveys** can do much to supplement the sources of data discussed above. An example from Britain is the International Passenger Survey, a sample survey carried out at seaports and airports. It was established to provide information on tourism and the effect of travel expenditure on the balance of payments, but it also provides useful information on international migration.

However, even when all the available sources of information are used to analyse migration patterns there is no doubt that a large proportion of population movements go entirely unrecorded.

Now test yourself

Tested

6 Which American economist is associated with the cost–benefit approach to migration?
7 List the three components of the economic cost of migration.
8 Give **three** push factors relating to rural–urban migration in Brazil.
9 Give **three** pull factors relating to rural–urban migration in Brazil.
10 State the three principal sources of migration data.

Answers on p.216

5.2 Internal migration

Distance, direction and patterns ──────────── Revised ☐

Figure 5.4 provides a comprehensive classification of population movements, the majority of which are internal migrations.

As cost is a significant factor in the distance over which migration takes place, the relative distance of movements may have a filtering effect upon the kinds of people who are moving between different areas. There is also a broad relationship between social/cultural change and distance.

In terms of direction the most prevalent forms of migration in developing countries are from rural to urban environments and from peripheral regions to economic core regions.

Although of a lesser magnitude, rural–rural migration is common in the developing world for a variety of reasons including employment, family reunion and marriage. In some instances governments have encouraged the agricultural development of frontier areas such as the Amazon Basin in Brazil.

Distance	Direction
Intra-national	Rural–rural
Local	Rural–urban
Intra-district	Urban–rural
Inter-district	Urban–urban
Intra-provincial	Periphery–core
Inter-provincial	Core–periphery
Intra-regional	Traditional–modern spheres
Inter-regional	**Patterns**
International	Step migration
LEDCs–LEDCs	Migration stream
LEDCs–MEDCs	Counter-stream

Figure 5.4 Spatial dimensions of population movements in LEDCs

The causes of internal migration ──────────── Revised ☐

The macro-level dimension

This dimension highlights socio-economic differences at the national scale, focusing on the **core–periphery** concept. The development of core regions in many developing countries had its origins in the colonial era. At this time migration was encouraged to supply labour for new colonial enterprises and infrastructural projects such as the development of ports. In the post-colonial era most developing countries have looked to industrialisation as their path to development, resulting in disproportionate investment in the urban-industrial sector and the relative neglect of the rural economy. This has encouraged a high level of rural–urban migration.

The meso-level dimension

The meso-level dimension includes more detailed consideration of the factors at origin and destination that influence people's migration decisions. E. S. Lee's origin–intervening obstacles–destination model helps in understanding this level of approach, which recognises the vital role of the perception of the individual in the decision-making process.

High population growth is often cited as the major cause of rural–urban migration. However, in itself, population growth is not the main cause of out-migration. Its effects have to be seen in conjunction with the failure of other processes to provide adequately for the needs of growing rural communities. However, out-migration can provide an essential 'safety-valve' for hard-pressed rural communities. The economic motive underpins the majority of rural–urban movements. Michael Todaro was one of the first to recognise that the paradox of urban deprivation on the one hand and migration in pursuit of higher wages on the other could be explained by taking a long-term view of why people move to urban areas. Other factors, particularly the social environment, have a very strong influence on the direction that the movement takes.

The micro-level dimension

This level stresses that the specific circumstances of individual families and communities in terms of urban contact are of crucial importance in the decision to move, particularly when long distances are involved. The importance of established links between urban and rural areas frequently results in the

> The **core** is a region of concentrated economic development with advanced systems of infrastructure, resulting in high average income and relatively low unemployment.
>
> The **periphery** is a region of low or declining economic development characterised by low incomes, high unemployment, selective out-migration and poor infrastructure.
>
> **Macro-level** – large scale.
>
> **Meso-level** – intermediate scale.
>
> **Micro-level** – small scale.

phenomenon of **chain migration**. Apart from contact with, and knowledge of, urban locations, differentiation between rural households in terms of migration takes the following forms:

- level of income
- size of land holding
- size of household
- stage in the life cycle
- level of education
- cohesiveness of the family unit

> **Chain migration** occurs when one or a small number of pioneering migrants have led the way in rural–urban migration, and others from the same rural community follow.

Now test yourself

Tested

11 Give **three** reasons for rural–rural migration in the developing world.
12 List the three levels of scale that should be considered in a discussion of internal migration.
13 Define the terms (a) core, (b) periphery.
14 Which factor does the micro-level dimension of migration analysis suggest?

Answers on p.216

> **Expert tip**
>
> Where relevant you should always consider the concept of scale. It can be a very important organising factor in a discussion. Ask yourself: 'Is consideration of a range of scales relevant to this question?'

The impacts of internal migration

Revised

Socio-economic impact

Figure 5.5 illustrates the costs and returns from migration, highlighting the two-way transfers of labour, money, skills and attitudes. **Remittances** from internal migration are an important source of money for rural areas. However, the flow of money and support in general is not always one-way.

> **Remittance** is money sent home to families by migrants working elsewhere.

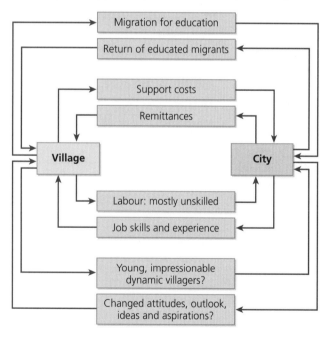

Figure 5.5 The costs and returns from migration

The relationship between migration and development is complex and still the subject of much debate:

- Migrants do move in reaction to newly developed opportunities.
- However, people in the poorest areas of developing countries do not exhibit the highest levels of out-migration – levels of literacy and skill may be so low that access to even very menial urban jobs may be difficult.
- Development in rural areas of origin often acts as a stimulus to out-migration, widening the horizons of a significant number among the rural population.

- Rural migrants are often the most dynamic young adults from their communities and should be of benefit to the receiving urban areas, providing enough opportunities are available for most to gain reasonable employment. However, newcomers can place a massive burden on over-stretched urban amenities and services, particularly if large numbers are unemployed.

The impact of out-migration on areas of origin is not at all clear. The traditional view has been that by reducing unemployment and underemployment, and providing inputs such as remittances and newly acquired skills, migration promotes development in rural areas of origin, narrows regional disparities and eventually makes migration unnecessary. However, such mobility may have an adverse effect on the economy of labour-exporting areas. An important issue is the impact of out-migration on local agriculture.

Political impact

Internal migration at a significant scale can have considerable political repercussions:

- Lower political representation occurs where migration results in **depopulation**.
- In contrast, where population is growing rapidly, partly at least as a result of in-migration, the political voice of such regions will become more important.
- Internal migration can significantly change the ethnic composition of a region or urban area, which may result in tension. In some countries governments have been accused of deliberately using internal migration to change the ethnic balance of a region. Tibet is an example where the in-migration of large numbers of Han Chinese has had a huge impact. Prior to the Chinese occupation of Tibet in 1950 very few Chinese lived in what is now the Tibetan Autonomous Region (TAR). This has changed completely, with Chinese migrants now in the majority in some parts of Tibet. Most Tibetans see this as an immense threat to the survival of their culture and identity.

Environmental impact

Large-scale rural–urban migration has led to the massive expansion of many urban areas in developing countries, which has swallowed up farmland, forests, floodplains and other areas of ecological importance. In turn, the increased impact of these enlarged urban areas is affecting environments even further afield in a variety of different ways. These include deforestation, demand for water and other resources, pollution, and the expansion of landfill sites.

Impact on population structures

The age-selective (and often gender-selective) nature of migration can have a very significant impact on both areas of origin and destination. Population pyramids for rural areas in developing countries frequently show the loss of young adults (and their children) and may also show a distinct difference between the number of males and females in the young adult age group due to a greater number of males than females leaving rural areas for urban destinations. However, in some rural areas female out-migration may be at a higher level than male out-migration. The reverse situation is frequently true for urban areas.

Urban–urban migration

Revised

Stepped migration

Figure 5.6 shows three ways **stepped migration** might occur in a developing country such as Nigeria. During the initial move from a rural environment to a relatively small urban area migrants may develop skills and increase their knowledge of and confidence in urban environments. They may become aware of better employment opportunities in larger urban areas and develop the personal contacts that can be so important in the migration process.

> **Typical mistake**
>
> While remittances from urban migrants to their families in rural areas are a very important source of income, the role of support costs from families in rural areas to new urban migrants is often forgotten.

> **Depopulation** is an absolute decline in the population of an area, frequently caused by out-migration.

> **Stepped migration** occurs when a rural migrant initially heads for a familiar small town and then after a period of time moves on to a larger urban settlement. Over many years the migrant may take a number of steps up the urban hierarchy.

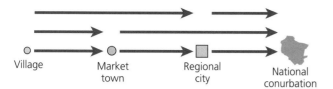

Figure 5.6 Stepped migration

Another important form of urban–urban migration is from towns and cities in economic periphery areas to urban areas in the economic core. An example is Brazil, with significant movement in the past 50 years from urban areas in the relatively poor northeast such as Recife and Salvador to the more prosperous cities of the southeast.

Causes and impacts of intra-urban movements

Demographic analysis shows that movements of population within cities are closely related to stages in the **family life cycle**, with the available housing stock being a major determinant of where people live at different stages in their life. Studies in Britain have highlighted the spatial contrasts in life cycle between middle- and low-income groups (Figure 5.7). While life cycle and income are the major determinants of where people live, residential patterns are also influenced by a range of organisations, including local authorities, housing associations, building societies and landowners. On top of this is the choice available to the household. For those on low income this is frequently very restricted indeed. As income rises the choice in terms of housing type and location increases.

> The **family life cycle** involves families with children passing through various stages over time (pre-child stage, family building, dispersal, post-child stage), with corresponding changes in housing needs.

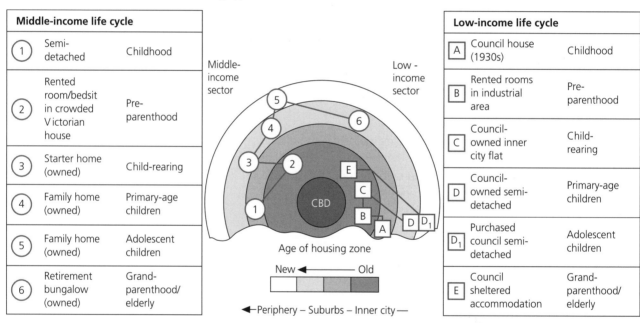

Housing choice is based on life cycle and income. Residential patterns are influenced by building societies, landowners, local authorities/housing associations, and free choice.

Source: *Advanced Geography: Concepts & Cases*, P. Guinness & G. Nagle (Hodder Education, 1999), p.104

Figure 5.7 Middle- and low-income models of the family life cycle in the UK

Counterurbanisation

Revised

Counterurbanisation first became clearly evident in the USA in the 1970s and since then most countries of Western Europe as well as Australia, New Zealand, Canada and Japan have followed suit. There has been much debate about the causes of counterurbanisation. The most plausible explanations are as follows:

- The 'period' explanation emphasises the role of the peculiar economic and demographic circumstances of the 1970s.

> **Counterurbanisation** is the process of population decentralisation as people move from large urban areas to smaller urban settlements and rural areas.

- The 'regional restructuring' explanation emphasises the role of the new organisation of production, the changing spatial division of labour and the increasing importance of service industries.
- The 'de-concentration' explanation highlights the lowering of institutional and technological barriers to rural location.

While all three explanations have their merits the third factor is viewed as the most important.

Now test yourself

Tested

15 With reference to Figure 5.5:
  (a) To what extent is rural–urban migration selective?
  (b) Why can 'support costs' flow from village to city.
16 What are remittances?
17 How can rural–urban migration impact on population structures?
18 Give an example of a country where stepped migration has occurred.
19 List the three explanations for counterurbanisation.

Answers on p.216

5.3 International migration

Voluntary and forced migration

Revised

Voluntary migration

International migration is a major global issue. In the past it has had a huge impact on both donor and receiving nations. In terms of the receiving countries the consequences have generally been beneficial. But today few countries favour a large influx of outsiders for a variety of reasons. Currently, one in every 35 people around the world is living outside the country of their birth. This amounts to about 175 million people, higher than ever before. Recent migration data show the following:

- With the growth in the importance of labour-related migration and international student mobility, migration has become increasingly temporary and circular in nature.
- The spatial impact of migration has spread, with an increasing number of countries affected either as points of origin or destination.
- The proportion of female migrants has steadily increased (now over 47% of all migrants).
- The great majority of international migrants from developed countries go to other affluent nations. Migration from less developed countries is more or less equally split between more and less developed countries (Figure 5.8). The movement between less developed countries is usually from weaker to stronger economies.
- Developed countries have reinforced controls, partly in response to security issues, but also to combat illegal immigration.

Globalisation has led to an increased awareness of opportunities in other countries. Each receiving country has its own sources of migration – the results of historical, economic and geographical relationships.

Forced migration

The abduction and transport of Africans to the Americas as slaves was the largest forced migration in history. In the seventeenth and eighteenth centuries 15 million people were shipped across the Atlantic Ocean as slaves. Even in

> **Voluntary migration** occurs when the individual or household has a free choice about whether to move or not.
>
> **Forced migration** occurs when the individual or household has little or no choice but to move.

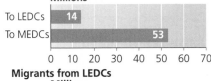

Migrants from MEDCs
Millions

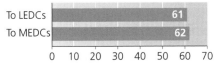

Migrants from LEDCs
Millions

Figure 5.8 Origin and destination of international migrants, 2005

recent times the scale of involuntary movement in the developing world is considerably higher than most people think.

In the latter part of the twentieth century and the beginning of the twenty-first century, some of the world's most violent and protracted conflicts have been in the developing world, particularly in Africa and Asia. These troubles have led to numerous population movements of a significant scale. Not all have crossed international frontiers to merit the term **refugee** movements. Instead many have involved **internal displacement**. This is a major global problem, which is showing little sign of abatement.

A number of trends appear to have contributed to the growing scale and speed of forced displacement:

- The emergence of new forms of warfare involving the destruction of whole social, economic and political systems.
- The spread of light weapons and land mines, available at prices that enable whole populations to be armed.
- The use of mass evictions and expulsions as a weapon of war and as a means of establishing culturally and ethnically homogeneous societies. The term 'ethnic cleansing' is commonly used to describe this process.

The United Nations High Commission for Refugees (UNHCR) is responsible for guaranteeing the security of refugees in the countries where they seek asylum and aiding the governments of these nations in this task.

Many parts of the developing world are prone to natural disasters. Because poor nations do not possess the funds to minimise the consequences of natural disasters as developed nations can, forced migration is often the result. Ecological and environmental change is a common cause of human displacement. Much of Central Asia is affected by problems such as soil degradation and desertification.

Increasingly large numbers of people have been displaced by major infrastructural projects and by the commercial sector's huge appetite for land. It is predicted that climate change will force mass migrations in the future. In 2009 the International Organisation for Migration estimated that worsening tropical storms, desert droughts and rising sea levels will displace 200 million people by 2050.

> A **refugee** is someone who has been forced to leave home and country because of 'a well-founded fear of persecution' on account of race, religion, social group or political opinion.
>
> An **internally displaced person** is someone who has been forced to leave his/her home for reasons similar to a refugee but who remains in the same country.

Typical mistake

Forced migration is not just the result of armed conflict, but can also occur due to environmental factors such as soil degradation and desertification.

Now test yourself

Tested

20 How many people around the world now live outside the country of their birth?

21 What is the difference between a refugee and an internally displaced person?

22 Suggest how climate change may cause forced migrations in the future.

Answers on p.216

The impacts of international migration

Revised

Socio-economic impact

Recent international migration reports have stressed the sharp rise in the number of people migrating to the world's richest countries for work. Such movement is outpacing family-related and humanitarian movements in many countries. The rise in labour-related migration has been for both temporary and permanent workers and across all employment categories. The distribution of immigrants in receiving countries is far from uniform, with significant concentration in economic core regions. Factors that influence the regional destination of immigrants into OECD countries are:

- the extent of economic opportunities
- the presence of family members or others of the same ethnic origin
- the point of entry into the country

The socio-economic status of OECD immigrants was frequently low. Immigrants were more likely to:

● be unemployed compared with nationals – in most European countries unemployment rates for foreigners are twice as high as for native workers
● have '3D' jobs that were 'dirty, dangerous and dull/difficult'
● be over-represented in construction, hospitality and catering, and in household services

The value of remittances has increased considerably over the past 20 years. The World Bank reports that international remittances totalled $397 billion in 2008, of which $305 billion went to developing countries. Some economists argue that remittances are the developing world's most effective source of financing. Although foreign direct investment is larger, it varies with global economic fluctuations. Remittances exceed considerably the amount of official aid received by LEDCs. Remittances have been described as 'globalisation bottom up'. Migration advocates stress that these revenue flows:

● help alleviate poverty – Figure 5.9 shows the relationship between remittances and poverty in Nepal
● spur investment
● cushion the impact of global recession when private capital flows decrease

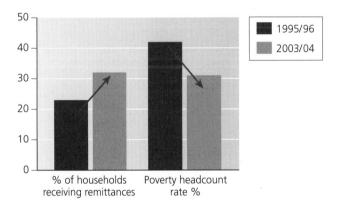

Figure 5.9 Remittances and poverty in Nepal

The major sources of remittances are the USA, Western Europe and the Persian Gulf. The top destinations of remittances are India, China, Mexico and the Philippines. The 20 million people who make up the Indian **diaspora** are scattered over 135 countries. In 2007 they sent back to India $27 billion. Remittances can form a positive **multiplier effect** in donor countries. In the past the perceived major disadvantage of emigration has been that it will lead to a 'brain drain' in which countries will lose their best workers. However, the direct and indirect effect of remittances may more than compensate for this.

> **Diaspora** refers to the dispersal of people from their original homeland.
>
> The **multiplier effect** is where an increase in the money supply in a region sets off an upward spiral of development as this money circulates in the economy.

The social assimilation of international migrants usually follows economic assimilation. Racial differences create the greatest barrier to social assimilation but differences in language, religion and culture can also be important. As social barriers decline the benefits that different cultures can bring to society as a whole become more apparent. The social impact on the donor country can also be considerable. This tends to occur in two stages. The first is the initial loss of many of its most dynamic individuals. The second stage occurs as new ideas from the adopted country filter back to the home country, often clashing with traditional values.

The cultural impact

Migration has played a major role in shaping the global cultural map. The old colonial powers have relatively cosmopolitan populations compared with most of their non-colonial counterparts as significant numbers of people from former colonies have sought a higher standard of living in the 'mother' country. Significant diaspora populations have been established in many developed countries, resulting in growing cultural hybridity.

A recent example is the enlargement of the European Union in 2004 to include Eastern European countries such as Poland. A considerable number of Polish workers migrated to the UK. In the USA the large inflow of migrants from Latin America has resulted in a substantial increase in the proportion of Spanish speakers in the country. Many areas in the southern part of the USA, in states such as California, New Mexico, Texas and Florida, are effectively bilingual.

The political impact

Significant levels of international migration can have a considerable political impact both within and between countries. In many countries there is a clear trend of immigrants being more likely to vote for parties of the centre and the left as opposed to political parties to the right of centre. In more developed countries immigrants tend to head for economic core regions and to inner city areas within these regions. Such concentrations can have a big impact on voting patterns.

Over time, immigrants gradually assimilate into host societies. In general, economic assimilation comes first followed by social assimilation and then political assimilation. When immigrant groups reach a certain size and standing they begin to develop their own politicians as opposed to voting for politicians from the host society.

High levels of international migration between one country and another can lead to political tension. The high level of Mexican migration into the USA, both legal and illegal, has created tensions between the US and Mexican governments. Many developing countries are looking to MEDCs to adopt a more favourable attitude to international migration. The subject is brought up regularly at international conferences. This political pressure is known as 'the pro-migration agenda of developing nations'.

The environmental impact

In an article entitled 'The environmental argument for reducing immigration to the United States', Staples and Cafaro explain how population growth contributes significantly to a host of environmental problems in the USA. They also argue that a growing population increases the USA's large environmental footprint beyond its borders and its disproportionate role in stressing global environmental systems. However, some critics see such arguments as a disingenuous way of attempting to curtail immigration.

> **Expert tip**
>
> Culture is the way of life of a particular society or group of people. Cultural traits or characteristics can be recognised in terms of language, customs, beliefs, dress, images, music, food and technology.

Case study | **Diasporas in London**

The diverse **ethnicity** of London is exemplified by the fact that over 200 languages are spoken within its boundaries. About two-thirds of immigration into Britain since the mid-1990s has been into London. Within the UK the process of **racial assimilation** is much more advanced in London than anywhere else. Almost 30% of people in London were born outside the UK, compared with 2.9% in northeast England.

The highest proportion of most ethnic groups in Britain can be found in one London borough or another. A range of factors affect ethnic concentration:
- There is a tendency for more recent immigrants to live in wards with a high ethnic minority concentration.
- Those who are not fluent in English are more likely to live in areas with a high ethnic minority concentration.
- Those in the highest social classes live in areas with a lower concentration of ethnic minority communities.

- Higher levels of qualification are associated with lower levels of ethnic minority concentration.
- The more paid workers there are in a household, the less likely they are to live in areas with a high ethnic minority concentration.

Ethnic villages

The concept of ethnic villages often appears in newspapers, magazines and academic journals. **Ethnic villages** show clear evidence of particular groups residing within their areas in terms of shops, places of worship, schools, cinemas, newspapers, social facilities, advertising and street presence. Ethnic villages in London include:
- Chinese in Soho
- Koreans in New Malden
- Bangladeshis in Tower Hamlets

Ethnicity is the identification of individuals with particular ethnic groups

Tested

Now test yourself

23 State **three** factors that influence the regional destination of immigrants in OECD countries.

24 How large were international remittances in 2008?

25 Define diaspora.

26 Describe the trends shown in Figure 5.9

27 State **three** factors that influence ethnic concentration in London.

Answers on p.216

5.4 A case study of international migration

Mexicans to the USA: a major migration stream

Revised

One of the largest international migration streams in the world in recent decades has been from Mexico to the USA. This has been primarily a **labour migration**, the result of a very large gap in:

- average income
- unemployment rates
- the growth of the labour force
- the overall quality of life

> **Labour migration** is migration from one country to another when the primary purpose is to seek employment.

About 30% of legal immigrants in the USA and an estimated half of all unauthorised foreigners in the country are from Mexico. Most of this migration has taken place in the last three decades. Table 5.2 summarises the main push and pull factors influencing migration from Mexico to the USA. Mexico is Latin America's major emigration country, sending up to 500,000 people — half of its net population increase to the USA each year. Most make unauthorised entries.

Table 5.2 Factors encouraging migration from Mexico, by type of migrant

Type of migrant	Demand-pull	Supply-push	Network/other
Economic	Labour recruitment (guest workers)	Unemployment or underemployment; low wages (farmers whose crops fail)	Job and wage information flows
Non-economic	Family unification (family members joins spouse)	Low income, poor quality of life, lack of opportunity	Communications; transport; assistance organisations; desire for new experience/adventure

Note: All three factors can encourage a person to migrate. The relative importance of pull, push and network factors can change over time.

> A **guest worker** is a foreigner who is permitted to work in a country on a temporary basis, for example a farm labourer.

Early and mid-twentieth century migration

In the early part of the twentieth century the US government allowed the recruitment of Mexican workers as **guest workers**. Young Mexican men known as 'Braceros' were allowed into the USA legally between 1917 and 1921, and then later between 1942 and 1964. Both guest worker programmes began when US farms faced a shortage of labour during periods of war. US farmers were strong supporters of the scheme, but trade unions were very much against it.

The increase in illegal migration

- There was very little illegal migration from Mexico to the USA before the 1980s.
- However, high population growth and the economic crisis in the early 1980s resulted in a considerable increase in illegal migration.

- Mexican migrants were employed mainly in agriculture, construction, various manufacturing industries and low-paid services jobs.
- As attitudes in the USA again hardened against illegal workers, Congress passed the Immigration Reform and Control Act (IRCA) in 1986. This imposed penalties on American employers who knowingly hired illegal workers.
- However, the Act also legalised 2.7 million unauthorised foreigners. Of this number, 85% were Mexican. The legalisation substantially expanded network links between Mexican workers and US employers.

2000 and beyond

The US Census in 2000 found an estimated 8.4 million unauthorised foreigners, who were mostly Mexican. This stimulated new attempts to regulate migration between the two countries. However, legal and illegal migration from Mexico continued as before. By 2006 there were an estimated 12 million Mexican-born people living in the USA. This amounted to around 11% of living people born in Mexico. With their children also taken into account the figure increased to more than 20 million. There is a very strong concentration of the US Mexican population in the four states along the Mexican border: California, Arizona, New Mexico and Texas. The concentration is particularly strong in California and Texas.

Mexican culture has had a sustained impact on many areas in the USA, particularly urban areas close to the border. However, there is also no doubt that the Mexican population in the USA has undergone a process of **assimilation** over time.

> **Assimilation** means becoming integrated into mainstream society.

The demography of Mexican migration to the US

With the US baby boom peaking in 1960, the number of US native-born people coming of working age actually declined in the 1980s. In contrast, high levels of fertility continued in Mexico in the 1960s and 1970s. The sharp increase in Mexico–US relative labour supply coincided with the stagnation of Mexico's economy in the 1980s, after significant economic progress in the 1960s and 1970s. This created ideal conditions for an emigration surge. However, with Mexico's labour supply growth converging with US levels, pressures for emigration from Mexico peaked in the late 1990s and are likely to fall in coming years.

> **Expert tip**
>
> Remember that assimilation occurs economically, socially and politically – and usually in this order.

Opposition to Mexican migration into the USA

The US Federation for American Immigration Reform (FAIR) argues that unskilled newcomers:
- undermine the employment opportunities of low-skilled US workers
- have negative environmental effects
- threaten established US cultural values

The recent global economic crisis saw unemployment in the USA rise to about 10% — the worst job situation for 25 years. Immigration always becomes a more sensitive issue in times of high unemployment. Those opposed to FAIR see its actions as uncharitable and arguably racist. Such individuals and groups highlight the advantages that Mexican and other migrant groups have brought to the country.

> **Typical mistake**
>
> Students at times make very generalised statements about attitudes to immigration. Try to be as specific as possible. For example, many employers favour a high rate of immigration because it increases the potential pool of labour, while trade unions sometimes oppose high immigration because this can keep wage rates lower than they would otherwise be.

The impact on Mexico

Sustained large-scale labour migration has had a range of impacts on Mexico. A **migrant culture** has become established in many Mexican communities. Significant impacts include:
- the high value of remittances, which totalled $25 billion in 2008
- reduced unemployment pressure, as migrants tend to leave areas where unemployment is particularly high

> **Migrant culture** refers to the attitudes and values of a particular society to the process of migration.

- lower pressure on housing stock and public services
- changes in population structure
- loss of skilled and enterprising people
- migrants returning to Mexico with changed values and attitudes

Now test yourself

Tested

28 State **four** differences between the USA and Mexico that have stimulated a high rate of migration from Mexico to the USA.

29 When did high illegal immigration from Mexico to the USA begin?

30 Which **two** states have the highest concentrations of people originating from Mexico?

Answers on p.216

Exam-style questions

Section A

Study Figure 5.4.

1 What are the main reasons for 'periphery–core migration'?   [2]

2 Suggest the possible adverse impact of high levels of rural–urban migration on rural areas in LEDCs.   [3]

3 What are the advantages and disadvantages of large in-migration into urban areas in LEDCs?   [5]

Section C

1 With reference to examples, distinguish between voluntary and forced movements.   [7]

2 Discuss the political barriers to international migration.   [8]

3 Examine the impact of one international migration stream on both the source and receiving areas.   [10]

Exam ready

6 Settlement dynamics

6.1 Changes in rural settlements

Rural settlements in both developed and developing countries have undergone considerable changes in recent decades for a number of reasons. These include:

- rural–urban migration
- urban–rural migration
- the consequences of urban growth
- technological change
- rural planning policies
- the balance of government funding between urban and rural areas

Changing rural environments in the UK   Revised ☐

The UK reflects many of the changes occurring in rural areas in other developed countries. In the past rural society was perceived to be distinctly different from urban society. However, there has been rapid rural change over the past 50 years or so:

- The economy of rural areas is no longer dominated by farming. As agricultural jobs have been lost, manufacturing, high technology and the service sector have increased.
- Other significant new users of rural space are recreation, tourism and environmental conservation.
- The **rural landscape** has evolved into a complex multiple-use resource and as this has happened the **rural population** has changed in character.

These economic changes have fuelled social change in the countryside with the in-migration of particular groups of people. In the post-war period the government has attempted to contain expansion into the countryside by creating **green belts** and by the allocation of housing to urban areas or to large key villages.

> The **rural landscape** is a mental or visual picture of countryside scenery, which is difficult to define, as rural areas are constantly changing and vary from place to place.
>
> The **rural population** comprises people living in the countryside in farms, isolated houses, hamlets and villages. Under some definitions small market towns are classed as rural.
>
> **Green belts** are areas of open land retained round a city or town over which there are wide-ranging planning restrictions on development.

Changing agriculture

The countryside has been affected by major structural changes in agricultural production. Although agricultural land forms 73% of the total land area of the UK, less than 2% of the total workforce is now employed in agriculture.

- The size of farms has steadily increased.
- Such changes have resulted in a significant loss of hedgerows, which provide important ecological networks.
- Agricultural wages are significantly below the national average and as a result farmers are among the poorest of the working poor.
- As many farmers have struggled to make a living from traditional agricultural practices, a growing number have sought to diversify.

> **Typical mistake**
>
> Students sometimes fail to distinguish clearly between different forms of rural settlement, such as villages and hamlets, both in terms of written description and identification on Ordnance Survey maps. Some definitions include small market towns as being rural and it is important to be able to distinguish these from larger urban entities.

> **Expert tip**
>
> Farm diversification means establishing sources of income beyond those of traditional farming. Examples are bed-and-breakfast accommodation and farm shops. Some farms diversify in only one way, while others diversify in a number of different ways.

Counterurbanisation and the rural landscape

In recent decades **counterurbanisation** has replaced urbanisation as the dominant force shaping settlement patterns. This has resulted in a 'rural population turnaround' in many areas where depopulation had been in progress.

Green belt restrictions have limited the impact of counterurbanisation in many areas adjacent to cities.

But, not surprisingly, the greatest impact of counterurbanisation has been just beyond green belts where commuting is clearly viable. Here rural settlements have grown substantially and been altered in character considerably.

Because of the geographical spread of counterurbanisation since the 1960s, the areas affected by **rural depopulation** have diminished. Depopulation is now generally confined to the most isolated areas of the country but exceptions can be found in other areas where economic conditions are particularly dire. Figure 6.1 is a simple model of the depopulation process.

> **Counterurbanisation** is the process of population decentralisation as people move from large urban areas to smaller urban settlements and rural areas.
>
> **Rural depopulation** is the decrease in population of rural areas, whether by out-migration or by falling birth rates as young people move away, usually to urban areas.

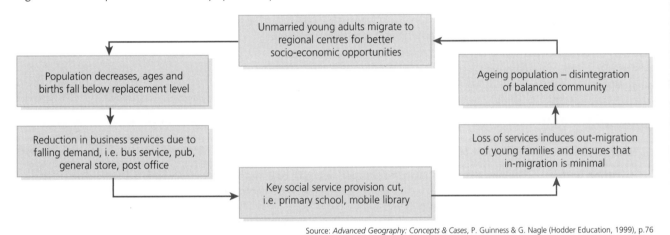

Source: *Advanced Geography: Concepts & Cases*, P. Guinness & G. Nagle (Hodder Education, 1999), p.76

Figure 6.1 Model of rural depopulation

The issue of rural services

Services are the basis for any community. Access to shops and post offices, healthcare and leisure activities create a feeling of belonging and a sustainable future for the area. However, rural services have been in decline for a number of decades, impacting heavily on the quality of life of many people, particularly those without a car. A 2008 report revealed that nearly half of communities had seen the loss of key local services in the previous four years.

ACRE (Action with Communities in Rural England) highlights the following reasons for rural service decline:

- The effect of market forces and, in some cases, the arrival of supermarkets in local areas making local services no longer competitive.
- The changing pattern of rural population, with more mobile residents with different shopping and consumer patterns becoming a greater part of the rural pattern of life.
- A change in expectations of rural residents themselves, who are no longer prepared to make do with relatively poor and expensive services and, in many cases, with the means and opportunity to access better services.

Key villages

Between the 1950s and 1970s the concept of key settlements was central to rural settlement policy in many parts of Britain, particularly where depopulation was occurring (Figure 6.2). The concept relates to central place theory and assumes that focusing services, facilities and employment in one selected settlement will satisfy the essential needs of the surrounding villages and hamlets. The argument was that, with falling demand, dispersed services would decline rapidly in vulnerable areas. The only way to maintain a reasonable level

> A **key village** is one designated as being one to develop in terms of the goods and services available to its own population and the population of the surrounding area.

of service provision in such an area was to focus on those locations with the greatest accessibility and the best combination of other advantages. In this way threshold populations could be assured and hopefully the downward spiral of service decline would be halted.

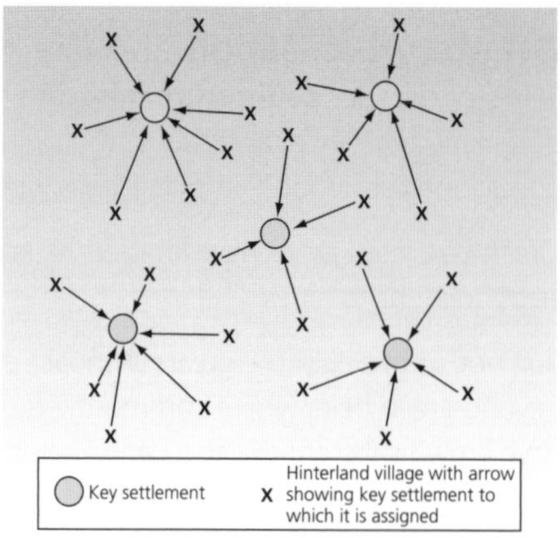

The rural transport problem

The considerable increase in car ownership in recent decades has had a devastating effect on public transport. While this has not disadvantaged rural car owners very much it has considerably increased the isolation of the poor, the elderly and the young.

The lack of public transport puts intense pressure on low-income households to own a car, a large additional expense that many could do without. Recent increases in the price of fuel have exacerbated this problem.

There has been continuing concern that Britain's remaining rural railway lines are under threat in a repeat of the 'Beeching cuts' of the 1960s.

The rural housing problem

The lack of affordable housing in village communities has resulted in a large number of young people having to move to market towns or larger urban centres. Only 12% of rural housing is subsidised, compared with 25% in urban areas. The issue of second homes has become increasingly contentious.

| ◯ Key settlement | **X** Hinterland village with arrow showing key settlement to which it is assigned |

Source: *Advanced Geography: Concepts & Cases*, P. Guinness & G. Nagle (Hodder Education, 1999), p.79

Figure 6.2 Key settlement concept

Case study The Isle of Purbeck

Table 6.1 Issues in rural settlement on the Isle of Purbeck

Location and development	The Isle of Purbeck forms the southeastern part of Purbeck District in Dorset. It is classed as a remote rural district. Here the rural settlement is concentrated in clustered villages with Corfe Castle being the largest. Although these villages are set in a network of isolated farms and houses, there are relatively few hamlets in the region. Lower-order urban services are provided by the towns of Swanage and Wareham, with higher-order urban services being found in the Bournemouth–Poole conurbation.
Population change	The population of Purbeck District as a whole has risen consistently over the past 40 years and is considerably older than that of the country as a whole, mainly because of the popularity of the area for retirement. However, the out-migration of young adults in search of wider economic opportunities and lower-cost accommodation is also a factor.
The rural housing problem	House prices in the area have risen at a rate above the national average. This has been due largely to competition from a number of different groups of people. This high level of competition for a limited number of available properties has pushed the cost of housing to a level well beyond the reach of most local people. The problem is compounded by the fact that local employment opportunities are limited and wages are low.
Rural service decline	The Dorset Rural Facilities Survey 2002 found a continuing decline in rural services. Service decline makes people more reliant on transport, both public and private, to gain access to basic services. The main causal factors have been increasing competition from urban supermarkets and the increasing personal mobility of most of the rural population.
Public transport	Public motor transport on the Isle of Purbeck is limited. There is extra minibus coverage through volunteer schemes, but this is also limited in extent.
Rural deprivation	Deprivation in terms of housing is particularly acute in high-priced housing counties such as Dorset. The lack of opportunity (opportunity deprivation) in health and social services, education and retail facilities also affects disadvantaged people, particularly those living in the most isolated rural areas. Mobility deprivation is also evident as public transport is very limited on the Isle of Purbeck. Deprivation is concentrated in the long-established population. Those who have migrated into the area generally have a significantly higher level of income.

Now test yourself

Tested

1 Define the term 'rural population'.
2 What proportion of the UK's workforce is now employed in agriculture?
3 What is counterurbanisation?
4 Give **two** reasons for rural service decline.
5 What is the main housing issue in rural areas?
6 **(a)** Where is the Isle of Purbeck?
   (b) Why have house prices risen at a rate above the national average?

Answers on p.216

Contemporary issues in rural settlements in LEDCs

Revised

The main process affecting rural settlements in developing countries has been rural–urban migration. In some areas it has provided a safety valve by:

- reducing rural population growth and pressure on food, water and other resources
- helping to limit unemployment and underemployment
- providing a valuable source of income through the remittances of migrants

However, in some rural communities the scale of rural–urban migration has been so great that it has resulted in:

- rural depopulation and an ageing population
- the closure of services, both public and private, as population declines
- insufficient labour to maintain agricultural production at its former levels

In southern African countries such as Botswana the devastating impact of AIDS has resulted in rural depopulation in many areas.

Rural poverty accounts for over 60% of poverty worldwide, reaching 90% in some developing countries like Bangladesh. In almost all countries access to education, health care, potable water, sanitation, housing, transport and communication is far worse than for the urban poor. Much urban poverty is created by the rural poor's efforts to get out of poverty by moving to cities.

About a third of the population of Mongolia live as nomadic herders on sparsely populated grasslands. Most live in very isolated locations. In recent years, droughts and unusually cold and snowy winters have decimated livestock, destroying the livelihoods of hundreds of thousands of households. Many have moved to Ulaanbaatar, the capital city, where they live in impoverished conditions, mainly on the periphery of the city. This exemplifies the concept of the **urbanisation of poverty**.

> **Urbanisation of poverty** is the increasing concentration of poverty in urban areas in developing countries due at least partly to high levels of rural–urban migration.

Now test yourself

Tested

7 What proportion of poverty worldwide is accounted for by rural poverty?
8 What is the 'urbanisation of poverty'?

Answers on p.216

6.2 Urban trends and issues of urbanisation

Urban trends

The first cities and the urban industrial revolution

Gordon Childe used the term **urban revolution** to describe the change in society marked by the emergence of the first cities some 5,500 years ago. The second 'urban revolution' based on the introduction of mass production in factories commenced in Britain in the late eighteenth century. This was the era of the industrial revolution when industrialisation and urbanisation proceeded hand in hand. As the processes of the industrial revolution spread to other countries the pace of urbanisation quickened. By the beginning of the most recent stage of urban development in 1950, 27% of people lived in towns and cities, with the vast majority of urbanites still living in the developed world.

The Post-1945 urban 'explosion' in the developing world

Throughout history **urbanisation** and significant economic progress have tended to occur together. In contrast, the rapid urban growth of the developing world in the latter part of the twentieth century has in general far outpaced economic development. Because urban areas in the developing world have been growing much more quickly than the cities of the developed world did in the nineteenth century the term 'urban explosion' has been used to describe contemporary trends.

Current patterns

The most urbanised world regions are North America, Europe, Oceania and Latin America. The lowest levels of urbanisation are in Africa and Asia. In contrast, **urban growth** is highest in Asia and Africa, as these regions contain the fastest-growing urban areas. By 2025 half of the populations of Asia and Africa will live in urban areas and 80% of urban dwellers will live in the developing world. In the developed world levels of urbanisation peaked in the 1970s and have declined since then due to the process of counterurbanisation.

The cycle of urbanisation

The development of urban settlement in the modern period can be seen as a sequence of processes known as the **cycle of urbanisation**. The key processes are: **suburbanisation**, **counterurbanisation** and **reurbanisation**. In Britain suburbanisation was the dominant process until the 1960s. From then on counterurbanisation impacted increasingly on the landscape. Reurbanisation of some of the largest cities, beginning in the 1990s, is the most recent phenomenon.

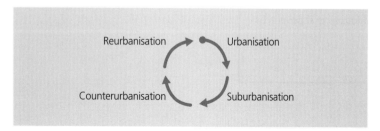

Figure 6.3 The cycle of urbanisation

The **urban revolution** was a major change in the form and growth of settlements due to significant technological advance.

Urbanisation is the process whereby an increasing proportion of the population in a geographical area lives in urban settlements.

Urban growth is the absolute increase in physical size and total population of urban areas.

Typical mistake

Students are often not clear about the difference between urbanisation and urban growth. Urbanisation involves an increasing proportion of a population living in urban areas. Urban growth can occur without this happening if urban areas are growing at a rate below the rate of population growth.

The **cycle of urbanisation** comprises the stages of urban change from the growth of a city to counterurbanisation through to reurbanisation.

Suburbanisation is the outward growth of towns and cities to engulf surrounding villages and rural areas.

Counterurbanisation is the process of population decentralisation as people move from large urban areas to smaller urban settlements and rural areas.

Reurbanisation occurs when, after a clear period of decline, the population of a city, in particular the inner area, begins to increase again.

Suburbanisation

In Britain the main factor in this development was the construction of suburban railway lines. Each railway development spurred a rapid period of house building. Initially the process of suburbanisation was an almost entirely middle-class phenomenon. It was not until after the First World War, with the growth of public housing, that working-class suburbs began to appear. The reasons for rapid suburban growth were:

- government support for house building
- the willingness of local authorities to provide piped water, sewerage systems, gas and electricity
- the expansion of building societies
- low interest rates
- development of public transport routes
- improvements to the road network

Counterurbanisation

Urban deconcentration is the most consistent and dominant feature of population movement in most developed cities today, in which each level of the settlement hierarchy is gaining people from the more urban tiers above it but losing population to those below it. However, it must be remembered that the net figures hide the fact that there are reasonable numbers of people moving in the opposite direction.

Reurbanisation

In very recent years British cities have, to a limited extent so far, reversed their previous population decline. Central government finance, for example the millions of pounds of subsidies poured into London's docklands, has been an important factor in the revival. New urban design is also playing a role.

For the first time in about 30 years London stopped losing population in the mid-1980s and has been gaining people ever since, due to net immigration from overseas and natural increase. Perhaps the most surprising aspect of this trend is the rejuvenation of inner London. Young adults now form the predominant population group in inner London, whereas in the 1960s all the inner London boroughs exhibited a mature population structure. Inner London is seen as a vibrant and attractive destination by young migrants from both the UK and abroad.

Issues of urbanisation
Revised

Competition for land

All urban areas exhibit competition for land, which varies according to location. The best measures of competition are the price of land and the cost of rents. However, planning measures such as **land use zoning** and other restrictions can complicate the free market process to a considerable degree. There are areas of some cities where dereliction has been long-standing. Here the land may be unattractive for development and it may require substantial investment to bring the area back into active use again.

> **Land use zoning** is a mapping exercise by local government which decides how land should be used in the various parts of a town or city.

Renewal and redevelopment

Urban redevelopment involves complete clearance of existing buildings and site infrastructure and construction of new buildings. In contrast, **urban renewal** keeps the best elements of the existing urban environment (often because they are safeguarded by planning regulations) and adapts them to new usages.

In cities where damage was extensive as a result of the Second World War, large-scale redevelopment took place in the following decades. The general model was to completely clear the land (redevelopment) and build anew. However, from the 1970s renewal gained increasing acceptance and importance in planning circles. In more recent years the term **urban regeneration** has become increasingly popular. This involves both redevelopment and renewal.

In the UK, urban development corporations were formed in the 1980s and early 1990s to tackle large areas of urban blight in major cities. The establishment of the London Docklands Development Corporation in 1981 set in train one of the largest urban regeneration projects ever undertaken in Europe. An important part of this development was the construction of Canary Wharf, which extended London's CBD towards the east. The regeneration of the Lower Lea Valley in preparation for the 2012 Olympic Games is a recent example. It is hoped that the resulting transformation of the area will bring permanent prosperity through the process of **cumulative causation**.

> **Expert tip**
>
> Remember that urban regeneration is the overall term for large-scale change and improvement of the urban landscape. It can just involve urban redevelopment, but might also contain aspects of urban renewal.

> **Cumulative causation** is the process whereby impulses for economic growth are self-reinforcing, resulting in an upward spiral of economic development.

Now test yourself

Tested

9 What is the difference between urbanisation and urban growth?

10 List the three processes in the cycle of urbanisation.

11 Define land use zoning.

12 Distinguish between urban redevelopment and urban renewal.

13 Regeneration in which part of London has benefited due to investment for the 2012 Olympic Games?

Answers on pp.216–217

Gentrification: reshaping social geography

There are two main reasons for clusters of high socio-economic status in the inner city:

- Some areas have always been fashionable. In London areas such as St John's Wood and Chelsea are only a short journey to the CBD, and pleasantly laid out with a good measure of open space. The original high quality of housing has been maintained to a very good standard.
- Other fashionable areas have become so in recent decades through the process of **gentrification**.

> **Gentrification** is a process in which wealthier people move into, renovate and restore run-down housing in an inner city or other neglected area. Such housing was formerly inhabited by low-income groups, with the tenure shifting from private-rented to owner-occupied.

The term gentrification was coined in 1963 by the sociologist Ruth Glass to describe the changes occurring in the social structure and housing market in parts of inner London such as Paddington and Fulham. The low-income areas that are most likely to undergo this process usually have some distinct advantages, such as:

- an attractive park
- larger-than-average housing
- proximity to a station

Evidence of gentrification includes:

- many houses being renovated
- house prices rising faster than in comparable areas
- 'trendier' shops and restaurants opening in the area

Because the demand for housing in London exceeds the supply, many parts of the city have been gentrified since the 1960s.

Changing accessibility and lifestyle

As cities spread outwards with the development of new suburbs, people enjoyed a higher quality of life in such locations. For many this resulted in longer journeys to work, but most could afford to spend money on transportation, and levels of **accessibility** rose.

> **Accessibility** refers to the relative ease with which a place can be reached from other locations.

However, significant investment in transport infrastructure is required to ensure that congestion does not reduce travel times. Reasons for the increase in urban car use in most cities include:

- rising real incomes, which have enabled more people to buy cars
- decentralisation, which has resulted in people living further from their places of work
- the growth in the number of households, which has generated more trips
- a growing proportion of working-age families having two earners, which also generates more trips
- the perceived high cost and low quality of public transportation, which limits its appeal as an alternative to the car
- the increasing proportion of journeys to school taken by car

The fastest rates of motor vehicle increase are in the cities of the developing world. This is the major reason why the quality of the environment in most developing cities continues to deteriorate.

Global (world) cities

Figure 6.4 shows what are termed the 'alpha' **global cities** in 2008, which are subdivided into four categories.

> A **global (world) city** is a city that is judged to be an important nodal point in the global economic system.

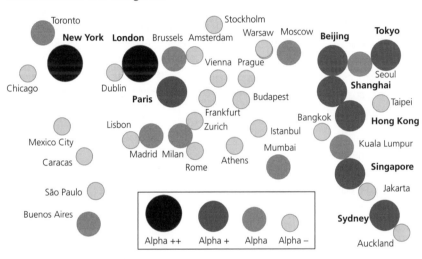

Figure 6.4 The world's alpha global cities

The growth of global cities has been due to:

- demographic trends – significant rates of natural increase and in-migration at different points in time for cities in the developed and developing worlds; large population clusters offer potential in terms of both workforce and markets
- economic development – the emergence of major manufacturing and service centres in national and continental space, along with the development of key transport nodes in the global trading system
- cultural/social status – the cultural facilities of large cities are an important element of their attraction to foreign direct investment and tourism
- political importance – many global cities are capital cities, benefiting from particularly high levels of investment in infrastructure

Now test yourself

Tested

14 When, and by whom, was the term gentrification first coined?
15 Define accessibility.
16 What is a global city?
17 Name the two 'alpha ++' cities shown in Figure 6.4.

Answers on p.217

6.3 The changing structure of urban settlements

Functional zonation

The concentric zone model
The main assumptions upon which the model was based are:

- a uniform land surface
- free competition for space
- universal access to a single-centred city
- continuing in-migration to the city, with development taking place outward from the central core

E. W. Burgess concluded that the city would tend to form a series of concentric zones (Figure 6.5), with the physical expansion of the city occurring by the processes of invasion and succession, with each of the concentric zones expanding at the expense of the one beyond.

Business activities agglomerated in the central business district (CBD), which was the point of maximum accessibility for the urban area as a whole. Surrounding the CBD was the '**zone in transition**' where older private houses were being subdivided into flats and bed-sitters or converted to offices and light industry. Newcomers to the city were attracted to this zone because of the concentration of relatively cheap, low-quality rented accommodation. In-migrants tended to group in ethnic ghettos.

Beyond the zone in transition came the 'zone of working-men's homes' characterised by some of the oldest housing in the city, and stable social groups. Next came the 'residential zone' occupied by the middle classes, with its newer and larger houses. Finally, the 'commuters' zone' extended beyond the built-up area.

> A **concentric zone** is a region of an urban area, circular in shape, surrounding the CBD and possibly other regions of a similar shape, that has common land use/socio-economic characteristics.
>
> The **zone in transition (twilight zone)** is the area just beyond the CBD, which is characterised by a mixture of residential, industrial and commercial land use, tending towards deterioration and blight. The poor quality and relatively cheap cost of accommodation makes this part of the urban area a focus for in-migrants, resulting in a rate of population change higher than in other parts of the urban area.

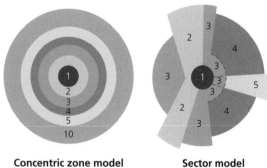

Concentric zone model

Sector model

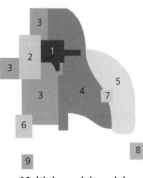

Multiple-nuclei model

1 Central business district
2 Wholesale light manufacturing
3 Low-class residential
4 Medium-class residential
5 High-class residential
6 Heavy manufacturing
7 Outlying business district
8 Residential suburb
9 Industrial suburb
10 Commuters' zone

Source: *OCR AS Geography* by M. Raw (Philip Allan Updates, 2008), p.203

Figure 6.5 Concentric zone, sector and multiple-nuclei models

The sector model

Hoyt placed the business district in a central location for the same reason – maximum accessibility (Figure 6.5). However, he observed that once variations arose in land uses near to the centre, they tended to persist as the city expanded. High-income housing usually developed where there were distinct physical or social attractions, with low-income housing confined to the most unfavourable locations. Major transport routes often played a key role in influencing sectoral growth, particularly with regard to industry. As new land was required by each sector it was developed at the periphery of that sector. However, medium- and high-class housing near the centre, the oldest housing in each case, was subject to suburban relocation by its residents, leading to deterioration, subdivision and occupation by low-income groups.

The multiple nuclei model

C. D. Harris and E. Ullman (1945) argued that the pattern of urban land use does not develop around a single centre but around a number of discrete nuclei (Figure 6.5). Some nuclei may be long established, for example old villages that have been incorporated into the city by urban expansion. Others, such as industrial estates for light manufacturing, are much newer. Similar activities group together, benefiting from agglomeration, while some land uses repel others. Middle- and high-income house buyers can afford to avoid residing close to industrial areas, which become the preserve of the poor. A very rapid rate of urban expansion may result in some activities being dispersed to new nuclei, such as an out-of-town shopping centre.

Bid-rent theory

Alonso's theory of urban land rent (1964), generally known as **bid-rent theory**, also produces a concentric zone formation, determined by the respective ability of land uses to pay the higher costs of a central location (Figure 6.6). The high accessibility of land at the centre, which is in short supply, results in intense competition among potential land users. The prospective land use willing and able to bid the most will gain the most central location. The land use able to bid the least will be relegated to the most peripheral location.

> **Expert tip**
>
> In modern urban analysis the zone in transition is the innermost part of the inner city. The inner city is located between the CBD and the suburbs.

> **A sector** is a section of an urban area in the shape of a wedge, beginning at the edge of the CBD and gradually widening to the periphery.

> **Bid-rent theory** refers to decreasing accessibility as you move out from the centre of an urban area, with corresponding declining land values, allowing (in theory) an ordering of land uses related to rent affordability.

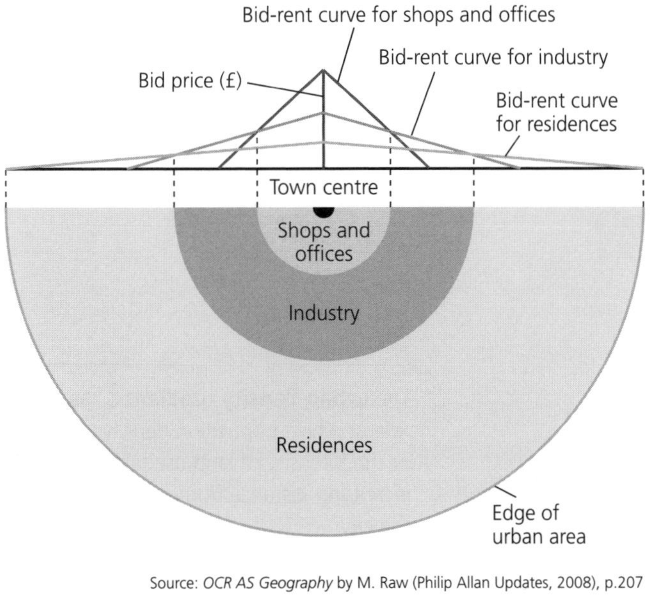

Source: *OCR AS Geography* by M. Raw (Philip Allan Updates, 2008), p.207

Figure 6.6 The bid-rent model

Models of developing cities

Griffin and Ford's model (Figure 6.7) summarises the characteristics of modern Latin American cities:

- Central areas that have changed radically from the colonial period to now exhibit most of the characteristics of modern western CBDs.
- The development of a commercial spine, extending outwards from the CBD.
- The tendency for industries with their need for urban services such as power and water to be near the central area.
- A 'zone of maturity', with a full range of services, containing both older, traditional-style housing and more recent development. The traditional housing has generally undergone subdivision and deterioration. A significant proportion of recent housing is self-built of permanent materials and of reasonable quality.
- A zone of 'in situ accretion', with varying housing types and quality, but with much still in the process of extension or improvement. Government housing projects are often a feature of this zone.
- A zone of squatter settlements, which is the place of residence of most recent in-migrants. Most housing is of the 'shanty' type.

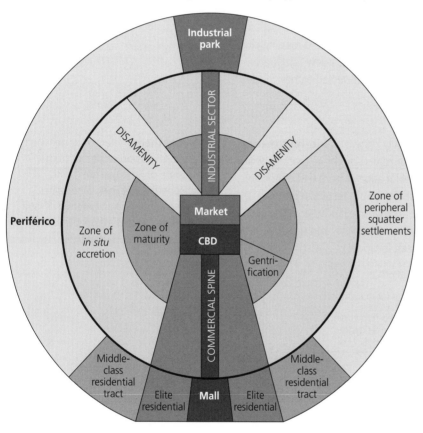

Figure 6.7 Latin American city model

Urban density gradients

Contrasting functional zones within urban areas characteristically vary in residential population density. Examination of population density gradients, termed gradient analysis, shows that for most cities densities fall with increasing distance from the centre. Gradient analysis of developed cities over time shows:

- the initial rise and later decline in density of the central area
- the outward spread of population and the consequent reduction in overall density gradient over time

In contrast, analysis of density gradients in developing countries shows:

- a continuing increase in central area densities
- the consequent maintenance of fairly stable density gradients as the urban area expands

> The **urban density gradient** is the rate at which population density and/or the intensity of land use falls off with increasing distance from the centre of the city.

Now test yourself

Tested

18 Who produced the concentric zone model?

19 In this model which area lies directly outside the CBD?

20 In both the concentric zone and sector models the CBD is centrally located due to which factor?

21 In the Griffin and Ford model where are the squatter settlements located?

22 Which term describes variations in population density from the centre to the periphery of a city?

Answers on p.217

Factors affecting the location of urban activities

Revised

A range of factors affect the location of urban activities such as retailing, manufacturing, as well as open space. Most can be placed under the general headings of:

- market forces – the demand and supply of land in various locations dictates its price
- local or central government planning decisions – planners can overrule market forces where they consider it necessary for the public good

Manufacturing industry

The first reaction to the constraints of inner city sites was to select new suburban locations, but increasingly manufacturing industry has been attracted to rural areas. The process of **deindustrialisation** has resulted in many factory closures in more affluent countries. The term **post-industrial city** is now commonly used in the developed world.

The explanation for the inner city decline of manufacturing industry lies largely in **constrained location theory**:

- The industrial buildings of the nineteenth and early twentieth century, mostly multi-storey, are generally unsuitable for modern manufacturing.
- The intensive nature of land use usually results in manufacturing sites being hemmed in by other land users, thus preventing on-site expansion.
- The size of most sites is limited and frequently deemed to be too small by modern standards.
- Where larger sites are available, the lack of environmental regulations in earlier times has often resulted in high levels of contamination, where reclamation is very costly indeed.
- The high level of competition for land in urban areas has continuously pushed up prices to prohibitive levels for manufacturing industry.

Although manufacturing employment has declined in cities as a whole, job losses have been much more severe in inner cities compared with suburban areas. Thus there has been a marked shift of manufacturing employment within urban areas in favour of the suburbs.

> **Deindustrialisation** is the long-term absolute decline of employment in manufacturing.
>
> A **post-industrial city** is a city whose economy is dominated by services and new high-tech industries.
>
> **Constrained location theory** identifies the problems encountered by manufacturing firms in congested cities, particularly in the inner areas.

Retailing

The location and characteristics of retailing have changed significantly in most cities. Outside the CBD large urban areas have witnessed the development of:

- suburban CBDs
- retail parks
- urban superstores
- out-of-town shopping centres
- internet shopping and home delivery services

Other services

The range of urban services that people use over a long time period can be extensive, often changing significantly during a person's lifetime. The location of some of these services may change more than others. For example:

● health

● education

● sport

Over the years an increasing number of land uses that require large sites and are mainly used by urban residents have been located in the **rural–urban fringe**.

> The **rural–urban fringe** is the boundary zone where rural and urban land uses meet. It is an area of transition from agricultural and other rural land uses to urban use.

The changing central business district

Most large CBDs exhibit a core and a frame (Figure 6.8). CBDs change over time. Common changes in many developed and an increasing number of developing countries have been:

● pedestrianised zones

● indoor shopping centres

● environmental improvements

● greater public transport coordination

● ring roads around the CBD, with multi-storey car parks

Some parts of the CBD may expand into the adjoining inner city (a zone of assimilation) while other parts of the CBD may be in decline (a zone of discard). The CBD is a major factor in the economic health of any urban area. Its prosperity can be threatened by a number of factors. CBDs are often in competition with their nearest neighbours and are constantly having to upgrade their facilities to remain attractive to their catchment populations. Urban redevelopment can be a major factor in CBD change.

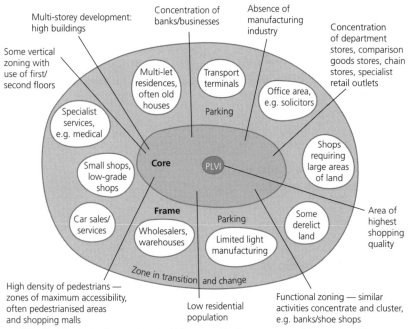

PLVI = Peak land value intersection: the highest rated, busiest, most accessible part of a CBD

Source: *AQA A2 Geography* by A. Barker, D. Redfern & M. Skinner (Philip Allan Updates, 2009), p.165

Figure 6.8 The key features of the CBD

Residential segregation

Residential segregation is very apparent in cities in both the developed and developing worlds. London provides a prime example. The contrast between the relative deprivation of inner London and the affluence of outer London is striking. The most intense deprivation in inner London is concentrated towards the east (the East End). The pattern found within boroughs is often quite intricate, forming the **residential mosaic** that social geographers frequently talk about. The process of gentrification invariably increases residential segregation.

6 Settlement dynamics

Typical mistake

Students can sometimes state generalisations as absolute fact. For example, although inner cities in general are considerably less affluent than the suburbs surrounding them, some parts of inner cities, particularly in large urban areas such as London, are extremely affluent, illustrating the concept of the residential mosaic. Notting Hill, Chelsea and St John's Wood in London are examples of inner city affluence.

> The **residential mosaic** is the complex pattern of different residential areas within a city reflecting variations in socio-economic status that are mainly attributable to income, but also influenced by ethnicity and age.

Now test yourself   `Tested`

23 Define deindustrialisation.

24 State **two** elements of constrained location theory.

25 Name the two sub-sections of the CBD.

26 What is the residential mosaic?

Answers on p.217

6.4 The management of urban settlements

Squatter settlements in São Paulo   `Revised`

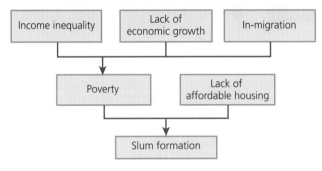

Figure 6.9 Inequality, poverty and slum formation

The slum housing problem

The population of São Paulo's metropolitan area (2000 census) is almost 18 million. At approximately 8110/km² the population density is more than three times that of Los Angeles. It is estimated that substandard housing occupies 70% of São Paulo's area:

- Two million people, 20% of the population, live in *favelas*.
- Over half a million people live in converted older homes in São Paulo's inner core, which are known as *cortiços*.

Heliopolis is São Paulo's largest slum. One hundred thousand people live here in a mix of absolute and semi-poverty. Squatter settlements are located:

- near gullies
- on floodplains
- on river banks
- along railways
- beside main roads
- adjacent to industrial areas

> **Slums** are run-down areas of a city characterised by sub-standard housing and squalor and lacking in tenure security.
>
> *Favelas* is a Brazilian term for informal, shanty-type settlements. They generally involve the illegal occupation of land by squatters.
>
> *Cortiços* comprise decaying formal housing, mainly in the inner city.

These are frequently areas that have been avoided by the formal building sector because of building difficulties and hazards. The nature of *favela* construction makes them vulnerable to fire, landslide and other hazards.

The transformation of *favelas*

Initially *favelas* are densely packed informal settlements made of wood, corrugated iron and other makeshift materials. Later they are replaced by concrete block construction. The large-scale improvements in *favelas* are due to residents' expectations of remaining where they are as a result of changes in public policy in the past 30 years from one of slum removal to one of slum upgrading.

Over time, a range of attempts have been made to tackle the housing crisis in São Paulo. These include:

- a federal bank funding urban housing projects and offering low-interest loans
- a state-level cooperatives institute that helped build housing for state workers

The administration of mayor Luiza Erundina (1989–1992) tried to speed up public house building. Here the emphasis was on **self-help housing initiatives**, known as *mutiroes*. The city supplied funding directly to community groups. The latter engaged local families in building new houses, or renovating existing ones.

Projeto Cingapura (the Singapore Project)

This urban renewal plan, based on the experience of Singapore, ran from 1995 to 2001. It was abandoned after it had provided only a modest increase in the available housing stock. At the outset, São Paulo's planners felt that the Singapore model was especially applicable because of the limited availability and high cost of urban land in both cities. While there was general encouragement for the initiative, a range of problems resulted in only 14,000 units being constructed as opposed to the 100,000 planned.

> **Typical mistake**
>
> Students often state that *favelas* are only located on the edge of developing cities. Often the largest ones are, but they tend to develop on any available land anywhere in developing cities.

> **Self-help housing initiatives** are partnerships between communities and local government whereby local government frequently supplies building materials and the community supplies the labour.

> ## Now test yourself   Tested
>
> 27 What is the population and population density of São Paulo?
> 28 Name São Paulo's largest slum.
> 29 Name the housing project that ran from 1995 to 2001.
>
> **Answers on p.217**

The provision of infrastructure for a city: Cairo   Revised

Cairo is the largest city in Africa. The population has risen rapidly over the last 50 years (Figure 6.10). The average population density is about 30,000/km^2. Much of the infrastructure of Cairo is designed for a population of about 2 million people and thus is under considerable strain from the tremendous demands being put on it by a much larger population.

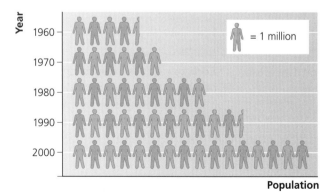

Figure 6.10 Cairo's population, 1960–2000

Transport

- Road transport has required a high level of investment as the city has expanded. Seven bridges now span the river Nile. Downtown expressways and flyovers bypass congested areas and new roads connect outlying suburbs. However, a 2007 survey of traffic congestion in Middle-East cities placed Cairo as the second most congested city after Dubai.
- Cairo's metro system ranks among the 15 busiest in the world with about 2 million passenger rides a day. Many informal settlements on the outskirts of Cairo are not adequately served by the public bus network or the metro.
- Cairo International is the second busiest airport in Africa. It has had to expand to keep up with demand.

Water supply and sanitation

- The Greater Cairo waste water project has provided a new sewerage system for Greater Cairo. Before the project was completed it was common to see sewage oozing from manholes in some parts of the city.
- In spite of considerable improvements, connections to the public sanitation network are missing from entire areas such as Batn El Ba'ara.
- Surface waters from the Nile river are the major source of bulk water supply in Cairo. However, its distribution system is inadequate. Significant new investment is required to improve this situation.

Housing

Three of the world's 30 largest slums are in Cairo. In total there are about 8 million slum dwellers in Greater Cairo. The poor find shelter in old graveyard tombs, mushrooming multi-storey informal settlements, sub-divided tenements, crumbling grand houses, new government apartment blocks and sub-standard inner-city houses. New urban initiatives such as the 10 new cities on the periphery of Cairo were designed to decentralise population and relieve the burden of in-migration, but the scheme has had only limited success. The housing problem remains immense.

Pollution infrastructure

Cairo suffers from high levels of air, noise and land-based pollution. The city is attempting to reduce the pollution problem:

- In 1998 36 monitoring stations were established to measure concentrations of sulfur and nitrogen dioxides, carbon monoxide, ozone and particulate matter.
- In 2004 the city acquired 70 rice-straw presses in order to reduce pollution caused by straw burning.
- Older buses are gradually being replaced by newer models with low-emission, natural-gas engines.

Now test yourself

Tested

30 Look at Figure 6.10. What was the population of Cairo in 2000?
31 Name the major project that improved the sewerage system.
32 How many new cities have been built on the periphery of Cairo?

Answers on p.217

> **Expert tip**
>
> It is useful to be able to distinguish between soft and hard infrastructure. The former covers housing, education health, leisure and other associated facilities. These are the social aspects of urban infrastructure. The latter refers to systems of transportation, communication, sewerage, water and electricity.

London: an MEDC inner city

Revised

There is a considerable gap in the **quality of life** between inner and outer London. Recently the population of inner London has increased (reurbanisation) after a long period of decline (Figure 6.11).

> **Quality of life** sums up all the factors that affect a person's general well-being and happiness.

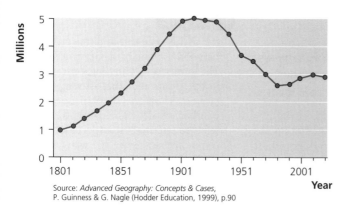

Source: *Advanced Geography: Concepts & Cases*,
P. Guinness & G. Nagle (Hodder Education, 1999), p.90

Figure 6.11 Population change in inner London

Characteristics of inner London

● Half of all adults are single, compared with 30% across England as a whole.

● 48% of the population of inner London are aged between 20 and 44, compared with the national average of 35%.

● In inner London over half of workers travel to work by public transport and less than a quarter by motor vehicle.

● Inner London has the lowest proportion of owner-occupiers in England, at 38%. The national average is 68%.

The 2004 Index of Multiple **Deprivation** showed that inner London boroughs such as Tower Hamlets, Hackney and Lambeth figured heavily in the 50 most deprived districts in the country.

> **Deprivation** is defined by the Department of the Environment as when 'an individual's well-being falls below a level generally regarded as a reasonable minimum for Britain today'.

The inner city problem and solutions

The nature of, and linkage between, London's inner-city problems is reasonably illustrated by the web of decline, deprivation and despair (Figure 6.12).

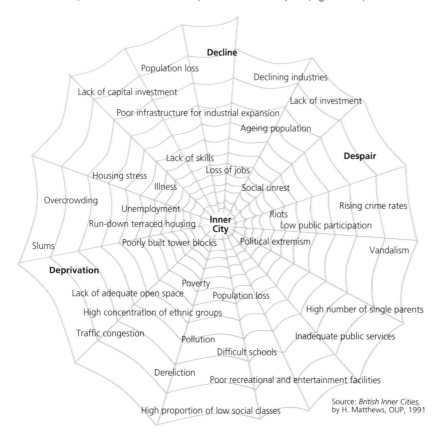

Source: *British Inner Cities*,
by H. Matthews, OUP, 1991

Figure 6.12 Inner-city web of decline, deprivation and despair

- Although not all local-authority housing estates are in inner cities, a significant proportion are.
- Recent studies have confirmed that such estates are increasingly becoming the focal points of deprivation, with very high levels of unemployment.
- Half of all council tenants are now in the poorest fifth of the population and those entering such housing are far poorer than those moving out.
- The processes of reurbanisation and gentrification are likely to accelerate this division.

Government responses since the late 1990s have focused on trying to tackle the **social exclusion** of the poor by targeting health, education, employment and child poverty in particular.

Two large-scale regeneration projects merit special attention – the redevelopment of London Docklands, which began in the 1980s, and the regeneration of the Lower Lea Valley in association with the 2012 Olympic Games. A range of smaller-scale projects have transformed the landscape in many inner city areas, but very significant problems still remain.

> **Social exclusion** is the process whereby certain groups are pushed to the margins of society and prevented from participating fully by virtue of their poverty, low education or inadequate life skills. This distances them from job, income and education opportunities as well as social and community networks.

Transport and pollution

London's air quality remains the worst in the UK. Particularly high pollution levels are recorded in large parts of central and inner London, while high levels of congestion and pollution have been caused by increasingly high volumes of motor vehicles. Road traffic accounts for 60% of nitrogen oxide emissions and 70% of fine particulate emissions in London. The establishment of the Congestion Charge zone and the Low Emissions zone are helping to tackle this problem.

> **Now test yourself**
>
> 33 When did the population of inner London peak (Figure 6.11)?
> 34 What proportion of households in inner London are owner-occupiers?
> 35 Since the late 1990s government policies to tackle poverty have focused on which phenomenon?
>
> **Answers on p.217**
>
> Tested

China: strategies for reducing urbanisation

Revised

Strategies for reducing urbanisation and urban growth in developing countries include:

- encouraging fertility decline
- promoting agricultural development in rural areas
- providing incentives to companies to relocate from urban to rural areas
- providing incentives to businesses to develop in rural areas
- developing the infrastructure of rural areas

Restricting rural–urban migration: China's Hukou system

The rate of urbanisation has been rising steadily in China (Figure 6.13).

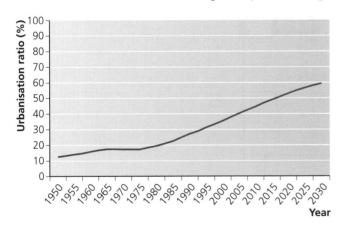

Figure 6.13 Urbanisation in China

From the 1950s the main instrument used to control rural–urban migration was the Hukou system, a population register that identified people as either 'urban' or 'rural'. Permission was required to leave the countryside and was only given if potential migrants could produce documentary evidence that they had an urban job to go to.

Alongside these measures, the authorities since the 1950s have periodically 'encouraged' large numbers of people to leave the cities. The 'back to the villages' movement in the early 1960s saw 20 million people leave large cities to return to their rural origins.

A relaxation of controls on rural–urban migration in the 1980s resulted in rapid urbanisation. However, the household registration system continues to provide the framework within which migration takes place. Local authorities in rural areas continue their efforts to limit out-migration, while local governments in city destinations have erected barriers in terms of employment discrimination and the 'deportation' of migrants back to their areas of origin.

Balanced development

China's urbanisation strategy emphasises balanced development. Large-scale urbanisation leads to an influx of rural population to the urban areas, bringing immense pressure to the large cities. In order to reduce urbanisation's impact on the large cities, China's urbanisation strategy focuses on promoting the development of small and medium-sized cities, ensuring multi-faceted development of regional economies.

In situ urbanisation

The emergence and development of **in situ urbanisation** has been one of the major characteristics of urbanisation in China since the 1980s. Over 20,000 small towns in China have developed in this way. The advantages of this process are seen as:

● benefiting significant numbers of the rural population, who are often neglected in the development process

● diverting many potential rural–urban migrants, who would otherwise head for slum areas in established cities

In situ urbanisation is well developed in Quanzhou in Fujian Province. Significant economic development in recent decades has seen the proportion of workers employed in agriculture fall considerably. Rather than moving to existing cities, especially large ones, most of Quanzhou's rural dwellers have been absorbed by township and village enterprises (TVEs).

> **In situ urbanisation** occurs when rural settlements transform themselves into urban or quasi-urban entities, with very little movement of population.

Now test yourself

Tested ▢

36 What is the name of China's population registration system?

37 What is the name given to the process of promoting the development of small and medium-sized cities?

38 Define in situ urbanisation.

Answers on p.217

Exam-style questions

Section A

1   Define rural depopulation.   [2]

2   Outline the reasons for rural depopulation.   [3]

3   In what ways can counterurbanisation change the character of rural settlements?   [5]

Section C

Study Figure 6.3.

1   Explain each of the processes in the cycle of urbanisation.   [7]

2   How can the competition for land in urban areas create distinct urban zones?   [8]

3   To what extent, and why, have the characteristics of central business districts changed in recent decades?   [10]

Exam ready

7 Tropical environments

7.1 Tropical climates

Characteristics of air masses
Revised

The concept of an air mass is generally applied only to the lower layers of the atmosphere. An air mass can cover an area up to tens of thousands of km^2.

Air masses derive their temperature and humidity from the regions over which they lie. These regions are known as source regions. The principle ones are:

- areas of relative calm, such as semi-permanent high-pressure regions
- areas where the surface is relatively uniform, including deserts, oceans and ice-fields

Air masses can be modified when they leave their sources.

A maritime tropical (mT) air mass refers to one that is warm and moist. A continental tropical (cT) air mass refers to one that is warm and dry.

As air masses move from their source regions they can be changed by the area over which they move. For example, a warm air mass that travels over a cold surface is cooled and becomes more stable. Hence, it may form low cloud or fog but is unlikely to produce much rain. By contrast, a cold air mass that passes over a warm surface is warmed and becomes less stable. The rising air is likely to produce more rain.

> An **air mass** is a large body of air in which the horizontal gradients of the main physical properties, such as temperature and humidity, are fairly gentle.

> **Now test yourself**
>
> 1 Define the term air mass.
> 2 Outline the main characteristics of mT and cT air masses.
>
> **Answers on p.217**
>
> Tested

Inter-tropical convergence zone (ITCZ)
Revised

The ITCZ or equatorial trough is a few hundred kilometres wide and lies at about 5°N. It wanders seasonally, lagging about 2 months behind the change in the overhead sun. The latitudinal variation is most pronounced over the Indian Ocean because of the large Asian continent to the north. Over the eastern Atlantic and eastern Pacific Oceans the ITCZ moves seasonally due to the cold Benguela and Peru currents.

Winds at the ITCZ are commonly light or non-existent, creating calm conditions called the doldrums, but there are occasional bursts of strong westerlies.

> The **inter-tropical convergence zone** is an area in the tropics in which the northeast trade winds and the southeast trade winds converge. If there is a difference in temperatures between the winds the warmer air rises over the denser, colder air and can produce rain.

Subtropical high pressure

- Centres or ridges of high pressure imply subsiding air. They tend to be found over continents, especially in winter.
- The subtropical high or warm anticyclone is caused by cold air at the tropopause descending. The position of the high pressure alters in response to the seasonal drift of the ITCZ.
- The subtropical high pressure belt tends to lie over the ocean, especially in summer, when there are low pressures over the continents caused by heating.
- Highs tend to be larger than low-pressure systems, reaching up to 4000 km in width and 2000 km north–south. Therefore, smaller pressure gradients are involved and so winds are lighter.

- The subtropical high generally moves eastwards at speeds of 30–50 km/hr. Hence a 4000 km system moving at an average of 40 km/hr would take about 4 days to pass over – if it kept moving.

Effects

- Where there is moisture at low level and air pollution, low-level stratus clouds may form, causing an anticyclonic gloom.
- Arid climates result from a prevalence of high pressure. Northeast Brazil is arid, even at a latitude of 8°S, because it protrudes far enough into the South Atlantic to be dominated by high pressure.

Now test yourself

3 Outline the main causes of the subtropical high-pressure belt.
4 How and why does it vary seasonally?

Answers on p.217

Tested

Oceanic influences

Revised

Ocean currents

The oceanic gyre (swirl of currents) explains why east coasts are usually warm and wet, because warm currents carry water polewards and raise the air temperature of maritime areas. In contrast, cold currents carry water towards the Equator and so lower the temperatures of coastal areas.

West coasts are cool and dry due to advection of cold water from the poles and cold upwelling currents.

Continental east coasts in the sub-tropics are humid and west coasts arid – this is mainly due to the easterly winds around the tropics.

Expert tip

The impact of an ocean current depends on its temperature – warm currents raise temperatures whereas cold currents tend to lower them.

Wind

The temperature of the wind is determined by the area where the wind originates and by the characteristics of the surface over which it subsequently blows. A wind blowing from the sea tends to be warmer in winter, but cooler in summer, than the corresponding wind blowing from the land.

Monsoon

The most famous monsoon occurs in India, but there are also monsoons in east Africa, Arabia, Australia and China. The most simple explanation for the monsoon is that it is a giant land–sea breeze:

- The great heating of the Asian continent and the high mountain barrier of the Himalayas, barring winds from the north, allows the equatorial rain systems to move as far north as 30°N in summer.
- In summer, central Asia becomes very hot, warm air rises and a centre of low pressure develops. The air over the Indian Ocean and Australia is colder, and therefore denser, and sets up an area of high pressure.
- As air moves from high pressure to low pressure, air is drawn into Asia from over the oceans. This moist air is responsible for the large amount of rainfall that occurs in the summer months.
- In the winter months the sun is overhead in the southern hemisphere. Australia is heated (forming an area of low pressure), whereas the intense cold over central Asia and Tibet causes high pressure. Thus in winter air flows outwards from Asia, bringing moist conditions to Australia.

This mechanism is, however, only part of the explanation.

The **monsoon** involves wind patterns that experience a pronounced seasonal reversal. The basic cause is the seasonal difference in heating of land and sea on a continental scale.

Now test yourself

5 Briefly explain the formation of the Asian monsoon.

Answer on p.217

Tested

Tropical climates

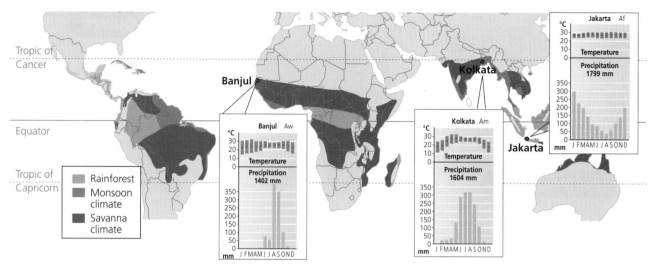

Figure 7.1 Tropical climates

The tropical environment is the area between 23.5°N and 23.5°S. This area covers about 50 million km² of land, almost half of it in Africa. According to the climatologist Koppen a tropical climate (A) is one in which the coldest month averages more than 18°C (the critical winter temperature for tropical forests). There are three types of tropical climate (Figure 7.1): rainforest (Af – no dry season), monsoon (Am – short dry season and heavy rains in the wet season) and savanna (Aw – a winter dry season).

Tropical humid climates have a mean monthly rainfall of over 50 mm for between 8 and 12 months. In contrast, **seasonally humid climates** have a mean monthly rainfall of over 50 mm for between 1 and 7 months.

Tropical humid climates (Af) are generally located within 5–10° of the Equator. Some higher latitudes may receive high levels of rainfall from unstable tropical easterlies. In Af climates the midday sun is always high in the sky – but high humidity and cloud cover keep temperatures from soaring. Some months, such as April and October, may be wetter due to movements of the ITCZ.

In contrast, seasonally humid climates (Aw) have a dry season, which generally increases with latitude. Rainfall varies from the moist low latitudes to semi-desert margins. As the midday sun reaches its highest point (zenith), temperatures increase, air pressure falls, and strong convection causes thundery storms. However, as the angle of the noon sun decreases, the rains gradually cease and drier air is re-established.

The tropical wet monsoon (Am) climate is a variation of the Aw climate. Winters are dry with hot temperatures. They reach a maximum just before the monsoon, and then they fall during the cloudy wet period when inflows of tropical maritime air bring high rainfall to windward slopes.

> **Tropical humid climates** refer to areas in the tropics that are wet all year round.
>
> **Seasonally humid climates** refer to areas within the tropics where there is a distinct dry season as well as a wet season.

Typical mistake

Many students forget about local factors when describing tropical climates. These include distance from the sea, altitude, aspect, ocean currents and human activities (land use).

Expert tip

When describing a climate graph you should refer to maximum and minimum temperatures, seasonal variations in temperatures, total rainfall, seasonal variations in rainfall and any links between temperatures and rainfall.

Now test yourself

6 Describe the main features of a savanna climate, as shown by the climate graph for Banjul in Figure 7.1.

Answer on p.217

7.2 Tropical ecosystems

Succession

Revised

- At the start of succession (pioneer **communities**), there are few nutrients and limited organic matter. Organisms that can survive are small and biodiversity is low.
- In late succession, there is more organic matter, higher biodiversity and longer-living organisms. Nutrients may be held by organisms, especially trees, so nutrient availability can be low.
- A well-documented case of succession is that following the eruption of Krakatoa in 1883.
- A contemporary example of succession being affected by arresting factors is the rainforest of Chances Peak in Montserrat.
- Studies in Puerto Rico showed that some forests recover from hurricane destruction in about 40 years.

Figure 7.2 shows a model of succession in a rainforest.

> **Succession** refers to the spatial and temporal changes in plant communities as they move towards a seral climax. Each **sere** or stage is an association or group of species that alters the micro-environment and allows another group of species to dominate.
>
> A **community** is a group of populations (animal and plant) living and interacting with each other in a common habitat (such as a savanna or a tropical rainforest).

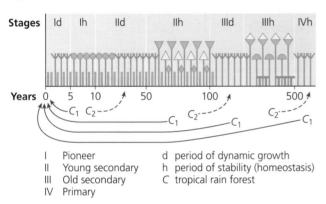

I	Pioneer	d	period of dynamic growth
II	Young secondary	h	period of stability (homeostasis)
III	Old secondary	C	tropical rain forest
IV	Primary		

Figure 7.2 A model of succession in a rainforest from bare soil to primary forest

The **climax community** is the group of species that are at a dynamic equilibrium with the prevailing environmental conditions. On a global scale, climate is the most important factor in determining large ecosystems or **biomes** such as tropical rainforest, and temperate woodland. In some areas, however, vegetation distribution may be determined by soils rather than by climate. This is known as **edaphic** control. For example, in savanna areas forests are found on clay soils, whereas grassland occupies sandy soils.

A **plagioclimax** refers to a plant community permanently influenced by human activity. It is prevented from reaching climatic climax by burning, grazing and so on. The maintenance of grasslands through burning is an example of plagioclimax.

Now test yourself

7 Define the term succession.
8 What is a plagioclimax?

Answers on p.217

Tested

Vegetation in tropical rainforests

Revised

The net primary productivity (NPP) of this ecosystem is 2,200 g/m^2/yr. This compares with the NPP for savanna of 900 g/m^2/yr, and agricultural land, 650 g/m^2/yr.

The hot, humid climate gives ideal conditions for plant growth and there are no real seasonal changes. Thus the plants are aseasonal and the trees shed their leaves throughout the year rather than in one season; the forest is turned evergreen as a result.

There is a great variety of plant species – in some parts of Brazil there are 300 tree species in an area of 2 km². The trees are tall and fast growing. The need for light means that only those trees that can grow rapidly and overshadow their competitors will succeed. Thus trees are notably tall and have long, thin trunks with a crown of leaves at the top; they also have buttress roots to support them. Trees are shallow rooted as they do not have problems getting water.

There are, broadly speaking, three main tiers of trees:

● **emergents**, which extend up to 45–50 m
● a **closed canopy** 25–30 m high, which cuts out most of the light from the rest of the vegetation and restricts its growth
● a limited **understorey** of trees, denser where the canopy is weaker; when the canopy is broken by trees falling, clearance or at rivers there is a much denser understorey

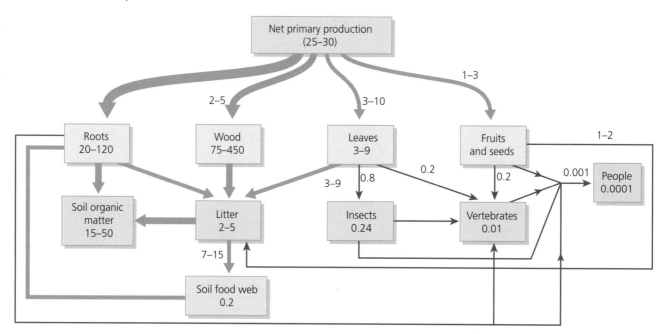

Figure 7.3 Energy flows in a tropical rainforest:

● The numbers inside the boxes represent dry matter of standing crops in tonnes/hectare.
● The numbers attached to the arrows represent flows in tonnes/hectare/year.
● The thickness of the arrows reflects the magnitude of the flows.

Now test yourself

9 Explain why tropical rainforests have very high rates of productivity.

Answer on p.217

Tested

Nutrient cycles

Revised

In the tropical rainforest, the input of nutrients from weathering and precipitation is high owing to the sustained warm, wet conditions. Most of the nutrients are held in the biomass due to the continual growing season. Breakdown of nutrients is rapid under the warm, wet conditions, and there is a relatively small store in the soil. Where vegetation has been removed, the loss of nutrients is enhanced due to high rates of leaching and overland flow.

In contrast, in a savanna grassland ecosystem the biomass store is less than that of the tropical rainforest because of the shorter growing season. The litter store is also small due to regular fires. This means that the soil store is relatively large.

Typical mistake

Many students think that tropical soils must be very fertile – in fact, they are generally very infertile, with the majority of nutrients being locked in the vegetation.

Now test yourself

Tested

10 Describe and explain the main characteristics of the nutrient cycle associated with tropical rainforests.
11 Outline the changes that occur as a result of human activity. Suggest reasons for the changes you have identified.

Answers on p.217

Savanna ecosystems

Savannas are areas of tropical grassland that can occur with or without trees and shrubs. There are a variety of types depending on soil type, drainage, geology and human factors, such as burning. The development and maintenance of savannas has occurred as a result of a variety of factors including climate, soils, geomorphology, fire and grazing (including domestic animals). Savannas are under increasing pressure from human activities. Trying to protect them is not proving easy.

Climate

The climate that characterises savanna areas is a tropical wet–dry climate. The wet season occurs in summer: heavy convectional rain (monsoonal) replenishes the parched vegetation and soil. Rainfall varies from as little as 500 mm to as much as 2000 mm – enough to support deciduous forest. However, all savanna areas have an annual drought. These can last from as little as 1 month to as much as 8 months.

Temperatures remain high throughout the year, ranging from 23°C to 28°C. The high temperatures, causing high evapotranspiration rates, and the seasonal nature of the rainfall cause division of the year into two seasons – one of water surplus and one of water deficiency.

Vegetation

Savanna vegetation is **xerophytic** (adapted to drought) and **pyrophytic** (adapted to fire):

- Adaptations to drought include deep tap roots to reach the water table, partial or total loss of leaves and sunken stomata on the leaves to reduce moisture loss.
- Adaptations to fire include very thick barks and thick budding that can resist burning, the bulk of the biomass being below ground level and rapid regeneration.

The warm, wet summers allow much photosynthesis and there is a large net primary productivity of 900 g/m^2/year.

Typical species in Africa include the acacia, palm and baobab trees, and elephant grass, which can grow to a height of over 5 m. Trees grow to a height of about 12 m and are characterised by flattened crowns and strong roots.

The role of fire, whether natural or man made, is very important. It helps to maintain the savanna as a grass community by mineralising the litter layer, killing off weeds, competitors and diseases, and preventing any trees from colonising relatively wet areas.

The fauna associated with savannas is very diverse. The African savanna has the largest variety of grazers and browsers – over 40 – including giraffe, zebra, gazelle, elephant and wildebeest. A variety of carnivores including lions, cheetahs and hyenas are also supported.

Soils

The soils of these areas are often heavily leached and **ferralitic**, with accumulations of residual insoluble minerals containing iron, aluminium and manganese. A distinction can be made between leached ferraltic soils (of the rainforest) and weathered ferralitic soils that are found in savanna regions. The hot, humid environment speeds up chemical weathering and decay of organic matter.

Not only are the soils well developed, but they have been weathered for a long time and are therefore lacking in nutrients and infertile. More than 80% of the soils are affected by acidity, low nutrient status, shallowness or poor drainage.

> **Expert tip**
>
> Give specific examples of vegetation and animals – typical savanna vegetation includes acacia, baobab and elephant grass, while mammals include giraffes, impala, wildebeest and zebra.

> **Now test yourself**
>
> 12 Outline the ways in which
> (a) savanna vegetation and
> (b) savanna fauna are adapted to seasonal drought.
> 13 Explain why fire is important in savanna ecosystems.
>
> **Answers on p.217**
>
> Tested

A typical soil catena for these areas can be identified, with ferruginous soils on the upper slopes, vertisols (tropical black clays) on the lower slopes, gleyed soils where drainage is impeded and alluvial soils close to rivers (Figure 7.4).

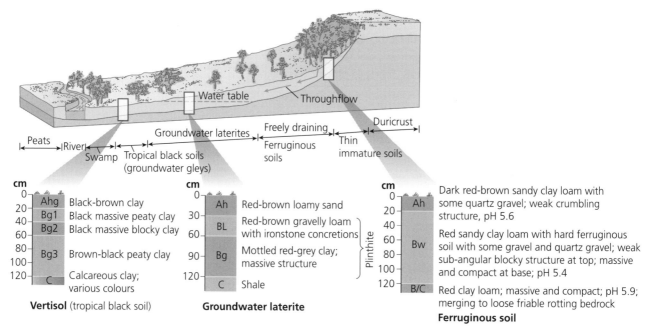

Figure 7.4 Soil catena in the savanna

7.3 Tropical landforms

Tropical landforms are diverse and complex. They are the result of many interrelated factors including climate, rock type, tectonics, time, vegetation, drainage, topography and, increasingly, human impact.

Weathering
Revised

Mechanical and chemical weathering occur widely in the tropics.

- In the humid tropics, the availability of water and the consistently high temperatures maximise the efficiency of chemical reactions, and in the oldest part of the tropics these have been operating for a very long period. In contrast, in many savanna areas, where there is less moisture, exfoliation or disintegration occurs.

- In many regions, weathering is complete and the **weathering profile** is very deep. As the depth of the weathered profile increases, slopes can become less stable. Rapid mass movements are likely to take place in a cyclical pattern, once a certain amount of weathering has occurred.

- Weathering profiles vary widely. The idealised weathering profile has three zones – residual soil, weathered rock and relatively unweathered bedrock. Weathered rock is also known as **saprolite**.

- In the weathered zone, at least 10% of the rock is unweathered **corestones**. This zone is typically highly permeable, especially in the upper sections, and contains minerals in a wide range of weathering stages.

- The 'weathering front' or '**basal surface of weathering**' between solid rock and saprolite (weathered rock) can be very irregular. Typically, deep weathering occurs to depths of 30–60 m, but because of variations in jointing density and rock composition, the depth varies widely over short distances.

Now test yourself

14 Describe the main types of weathering that occur in tropical environments.

Answer on p.217

Tested

Inselbergs, pediplains and etchplains

Tors

Most **tors** are found in strongly jointed rock. Tors vary in height from 20 m to 35 m and have core stones of up to 8 m diameter. They are formed by chemical weathering of the rock along joints and bedding planes beneath the surface. If the joints are widely spaced the core stones are large whereas if the joints are close together the amount of weathering increases and the corestones are much smaller.

Good examples of tors are found on the Jos Plateau of Nigeria and in the Matopas region and around Harare in Zimbabwe.

Inselbergs

Inselbergs are best developed on volcanic materials, especially granite and gneiss, with widely spaced joints and a high potassium content. These residual hills are the result of stripping weathered regolith from a differentially weathered surface.

The major debate is whether deep weathering is needed for hill formation. The two-stage model requires the development of a mass of weathered material beneath the ground and its subsequent removal. Alternatively, weathering and erosion could occur simultaneously. The diversity of residual hills suggests that both mechanisms operate simultaneously.

Monolithic domed inselbergs called **bornhardts** are characteristic landforms of granite plateaus of the African savanna, but can also be found in tropical humid regions. They are characterised by steep slopes and a convex upper slope. Bornhardts are eventually broken down into residual hills called castle kopjes.

Bornhardts occur in igneous and metamorphic rocks. Granite, an igneous rock, develops joints, up to 35 m below the surface, during the process of pressure release. Vertical jointing in granite is responsible for the formation of castle kopjes.

The two main theories for the formation of bornhardts include:
- the stripping or exhumation theory – increased removal of regolith occurs so that unweathered rocks beneath the surface are revealed
- parallel retreat, which states that the retreat of valley sides occurs until only remnant inselbergs are left

Classic examples of bornhardts include Mt Hora, in the Mzimba District of Malawi, and Mt Abuja in northern Nigeria.

Pediplains and etchplains

Pediplains are low-angled plains (pediments) separated by rocky hills known as kopjes. They are formed as a result of the parallel retreat of slopes (Figure 7.5). Pediplanation begins with tectonic uplift, resulting in accelerating river erosion forming knick points, falls, rapids and gorges along river valleys. When base level is reached, lateral erosion begins to occur.

> **Tors** are ridges or piles of spheroidically weathered boulders that have their bases in the bedrock and are surrounded by weathered debris.
>
> **Inselbergs** are isolated residual hills that stand prominently over a level surface.

Expert tip

Many features, such as tors and inselbergs, can be formed in different ways. Make sure you can give alternative explanations for their formation.

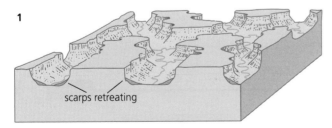
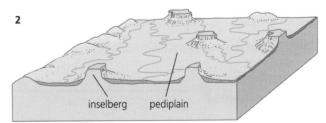

Figure 7.5 Pediplanation formed by parallel retreat

scarps retreating

inselberg   pediplain

An **etchplain** is an area of stripped and exposed unweathered bedrock (Figure 7.6). Etchplains occurs in ancient shield areas and are associated with deep weathering. Periods when weathering was more efficient resulted in the accumulation of great depths of weathered material. In contrast, there were periods when erosion was more rapid than weathering, leading to erosion of the weathered material, and exposure of the unweathered rock.

a The regolith surface is being lowered at a faster rate than the subterranean surface.

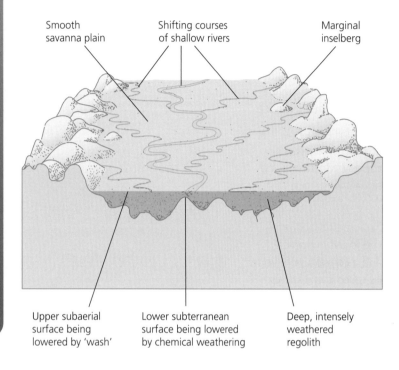

Smooth savanna plain

Shifting courses of shallow rivers

Marginal inselberg

Upper subaerial surface being lowered by 'wash'

Lower subterranean surface being lowered by chemical weathering

Deep, intensely weathered regolith

b The irregular subterranean surface is being exposed as inselbergs.

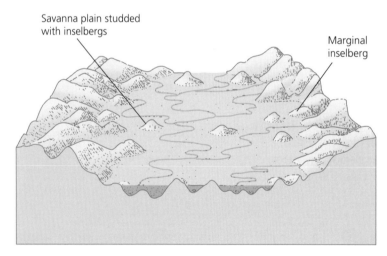

Savanna plain studded with inselbergs

Marginal inselberg

Figure 7.6 Etchplain formation

Tropical karst

There are two major landform features associated with tropical karst (Figure 7.7):

- **Polygonal** or **cockpit karst** is a landscape pitted with smooth-sided, soil-covered depressions and conical hills.
- **Tower karst** is a landscape characterised by upstanding rounded blocks set in a region of low relief.

Typical mistake

Some students consider that the types of limestone features found in temperate areas will be found in tropical areas – tropical limestone scenery is dominated by cockpit and tower karst.

Solution holes

The surface is broken up by many small solution holes but the overall surface remains generally level.

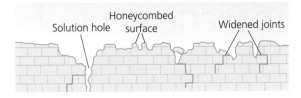

Cockpit karst

Cockpit karst is usually a hilly area in which many deep solution holes have developed to give it an 'eggbox' appearance.

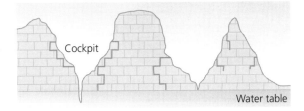

Tower karst

The widening and deepening of the cockpits has destroyed much of the limestone above the water table. Only a few limestone towers remain, sticking up from a flat plain of sediments that have filled in the cockpits at a level just above the water table. Eventually the towers will be entirely eroded, and disappear.

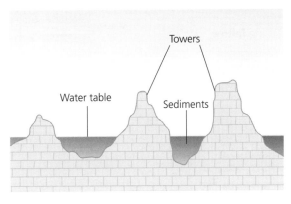

Figure 7.7 Cockpit karst and tower karst

Polygonal or cockpit karst is characterised by groups of hills, fairly uniform in height. These can be up to 160 m high in Jamaica, with a base of up to 300 m. They develop mainly as a result of solution. Polygonal karst tends to occur in areas that have been subjected to:

● high rates of tectonic uplift
● intense vertical erosion by rivers

The spacing of the hills may be related to the original stream network. Concentrated solution along preferred routes, such as wider joints, leads to accelerated weathering of certain sections of the limestone, especially during times of high flow. Water will continue to weather the limestone as far down as the water table. This creates closed depressions and dolines. Once the water table is reached, water will flow laterally rather than vertically, developing a flat plain.

By contrast, tower karst is much more variable in size than the conical hills of cockpit karst, with towers ranging from just a few metres to over 150 m in height in Sarawak. Other areas of tower karst include southern China, Malaysia, Indonesia and the Caribbean. The towers are characterised by steep sides, with cliffs and overhangs, and with caves and solution notches at their base. The steepest towers are found on massive, gently tilted limestone.

Where the water table is close to the surface, rivers will be able to maintain their flow over limestone, erode the surface and leave residual blocks set in a river plain. Other important processes include:

● differential solution along lines of weakness
● the retreat of cockpit karst slopes to produce isolated tower karst
● lateral erosion

Now test yourself

15 Explain why tors can be described as 'joint-controlled'.
16 Outline the two theories for the formation of bornhardts.
17 What is the difference between cockpit karst and tower karst?

Answers on p.218

Tested

7.4 Sustainable management of tropical environments

Sustainable agroforestry, Santa Rosa rainforest, Mexico

Revised

About 30,000 Popoluca Indians live in southern Mexico. The Popoluca of Santa Rosa farm the Mexican rainforest, using a form of agriculture known as the **milpa** system (Table 7.1), which resembles shifting cultivation but which mimics the natural rainforest:

● This is a labour-intensive form of agriculture, using **fallow**.
● It is a diverse form of **polyculture** with over 200 species cultivated, including maize, beans, cucurbits, papaya, squash, water melons, tomatoes, pineapples, chayotes, oregano, coffee and chili.
● Coffee is sold for cash.
● Two crops are planted annually.
● Fields are usually dug with digging sticks although a few households use ploughs.

> **Sustainable agroforestry** involves farming forests without destroying the natural environment or putting excessive strain on resources.

The variety of plants found in a natural rainforest is mirrored by using shifting cultivation of crops with different requirements. For example, lemon trees, peppervine and spearmint are **heliophytes** – light seeking – and prefer open conditions not shade. Coffee, by contrast, is a **sciophyte** – preferring shade – while the mango tree requires damp conditions.

The close associations that are found in natural conditions are also seen in the milpa system. For example, maize and beans go well together, as maize extracts nutrients from the soil whereas beans return them.

Tree trunks and small trees are left because they are useful for returning nutrients to the soil and preventing soil erosion. They are also used as a source of material for housing , hunting spears and medicines.

As in a rainforest, the crops are multi layered, with tree, shrub and herb layers. This increases NPP, because photosynthesis is taking place on at least three levels, and soil erosion is reduced, as no soil or space is left bare.

In all, 244 species of plant are used in the farming system. Most of the crops are self seeding, which reduces the cost of inputs. The Popolucas show a high level of ecological knowledge in managing the forest.

Table 7.1 A comparison between the milpa system and the new forms of agriculture

	Milpa system	Tobacco plantations or ranching
NPP	High, stable	Declining
Work (labour)	High	Higher and increasing
Inputs	Few	Very high: 2.5–3 tonnes fertiliser/ha/pa
Crops	Polyculture (244 species used)	Monoculture (risk of disease, poor yield, loss of demand and/or overproduction)
Yield (compared with inputs)	200%	140% if lucky
Reliability of farming system	Quite stable	High-risk operation
Economics	Mainly subsistence	Commercial
Income	None/little	More
Carrying capacity	Several families/4 ha plus livestock	1 family on a plantation (200 ha); ranching – 1 ha of good land/cow, 20 ha of poor land/cow

Animals

- Animals farmed include chickens, pigs and turkeys.
- These are used as a source of food, or for bartering and selling, and their waste is used as fertliser.
- Rivers and lakes are used for fishing and catching turtles.
- Deer, boar, rabbits and some birds are hunted with arrows and spears.
- It is not entirely a subsistence lifestyle since wood, fruit, turtles and other animals are traded for some seeds, mainly maize.

Pressures on the Popolucas

About 90% of Mexico's rainforest has been cut down in recent decades, largely for new forms of agriculture. This is partly a response to Mexico's huge international debt and attempts by the government to increase agricultural exports and reduce imports. The main new forms of farming are:

- **cattle ranching** for export
- **plantations** for cash crops, such as tobacco

However, these new methods are not necessarily suited to the physical and economic environment:

- Tobacco needs protection from too much sunlight and excess moisture and the soil needs to be very fertile.
- The cleared rainforest is frequently left bare and this leads to soil erosion. Unlike the milpa system, the new systems are very labour intensive.
- Pineapple, sugar cane and tobacco plantations require large inputs of fertiliser and pesticides. Inputs are expensive and the costs are rising rapidly.

Ranching prevents the natural succession of vegetation, because of a lack of seed from nearby forests and the grazing effects of cattle:

- Grasses and a few legumes become dominant.
- One hectare of rainforest supports about 200 species of trees and up to 10,000 individual plants. By contrast, one hectare of rangeland supports just one cow and one or two types of grass. But it is profitable, in the short-term, because land is available, and it is supported by Mexican government.

The Mexican rainforest can be described as a 'desert covered by trees'. Under natural conditions it is very dynamic, but its **resilience** depends on the level of disturbance.

> **Typical mistake**
>
> The Popolucas are not completely isolated – they trade and barter with other Mexicans. So they are not purely subsistence farmers.

Now test yourself

Tested

18 Compare the Popoluca's methods of farming with the natural tropical rainforest ecosystem. What lessons can be learnt from this?

19 The tropical rainforest is a 'desert covered by trees'. What does this mean?

Answers on p.218

Exam-style questions

1 (a) Compare and contrast the climate of humid tropical environments with that of seasonally humid environments. [10]

  (b) Explain how different tropical climates can affect human activities. [15]

2 (a) Outline the characteristics of vegetation in (i) the tropical rainforest and (ii) the savanna. In what ways are they adapted to their environments? [10]

  (b) Comment on the role of human activities in modifying the ecosystem of either tropical rainforests or savannas. [15]

Exam ready

8 Coastal environments

8.1 Waves, marine and sub-aerial processes

Features of waves

Revised

Waves result from friction between wind and the sea surface.

Wave height is an indication of wave energy. It is controlled by wind strength, fetch (the distance of open water a wave travels over) and the depth of the sea. Waves of up to 12–15 m are formed in open sea and can travel vast distances away from the generation area, reaching distant shores as **swell waves**, characterised by a lower height and a longer **wavelength**. In contrast, storm waves are characterised by a short wavelength, high amplitude and high **wave frequency**.

Wave height or **amplitude** is the distance between the trough and the crest.

Wavelength is the distance between two successive crests or troughs.

Wave frequency is the number of waves per minute.

Wave refraction

- As wave fronts approach the shore, their speed of approach will be reduced as the waves 'feel bottom'.
- Usually, wave fronts will approach the shore obliquely – this causes the wave fronts to bend and swing round in an attempt to break parallel to the shore.
- The change in speed and distortion of the wave fronts is called wave refraction (Figure 8.1).
- If refraction is completed, the fronts will break parallel to the shore.

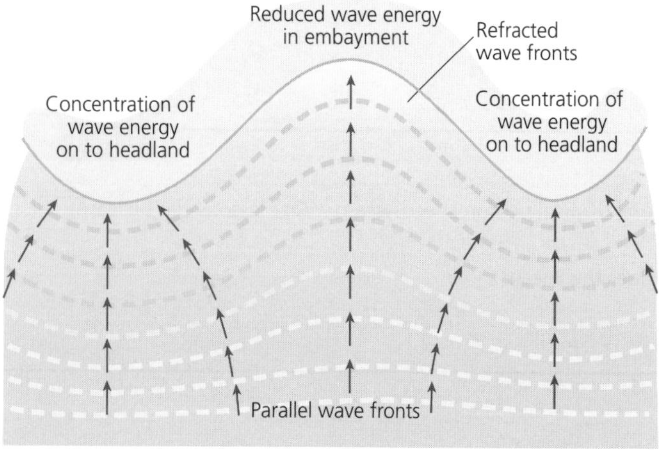

Source: *Advanced Geography: Concepts & Cases* by P. Guinness & G. Nagle (Hodder Education, 1999), p.294

Figure 8.1 Wave refraction

Wave refraction distributes wave energy along a stretch of coast. Along a complex transverse coast with alternating headlands and bays, wave refraction will concentrate wave energy and therefore erosional activity on the headlands, while wave energy will be dispersed in the bays. Hence deposition will tend to occur in the bays.

Breaking waves

- **Spilling breakers** are steep waves (large height relative to wavelength) associated with gentle beach gradients. They are characterised by a gradual peaking of the wave until the crest becomes unstable, resulting in a gentle spilling forward of the crest.
- **Plunging breakers** are waves of intermediate steepness that tend to occur on steeper beaches than spilling breakers. They are distinguished by the shoreward face of the wave becoming vertical, curling over and plunging forward and downward as an intact mass of water.
- **Surging breakers** have low steepness and are found on steep beaches. In surging breakers the front face and crest of the wave remain relatively smooth and the wave slides directly up the beach without breaking. In surging breakers a large proportion of the wave energy is reflected at the beach.

Once a breaker has collapsed, the **swash** will surge up the beach with its speed gradually lessened by friction and the uphill gradient. Gravity will then draw the water back as the **backwash**.

Constructive waves (Figure 8.2) tend to occur when wave frequency is low (6–8/minute), particularly when these waves advance over a gently shelving sea floor (e.g. formed of fine material, such as sand). Because of the low frequency, the backwash of each wave will be allowed to return to the sea before the next wave breaks – i.e. the swash of each wave is not impeded and retains maximum energy.

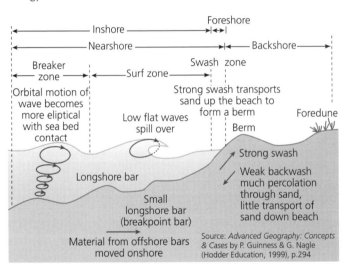

Figure 8.2 Constructive waves

Destructive waves (Figure 8.3) are the result of locally generated winds, which create waves of high frequency (12–14/minute). As the backwash is stronger than the swash, material is eroded from the beach.

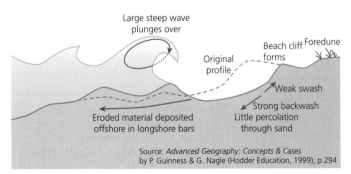

Figure 8.3 Destructive waves

> ## Now test yourself
>
> 1 Define wavelength and wave frequency.
> 2 Explain the meaning of the following terms: swash, backwash, fetch.
> 3 Distinguish between plunging breakers and surging breakers.
> 4 Describe the main changes that occur as a result of wave refraction.
>
> ### Answers on p.218
>
> Tested ☐

Coastal erosion

- **Hydraulic action (wave pounding)** is an important process as waves break onto cliffs (Figure 8.4). As the waves break against the cliff face, any air trapped in cracks, joints and bedding planes will be momentarily placed under very great pressure. As the wave retreats, this pressure will be released with explosive force.
- **Abrasion (corrasion)** is the process whereby a breaking wave can hurl pebbles and shingle against a coast, thereby abrading it.
- **Attrition** takes place when the eroded material itself is worn down.
- **Solution (corrosion)** is a form of chemical erosion which affects calcareous (lime-rich) rock. Waves speed up the process.

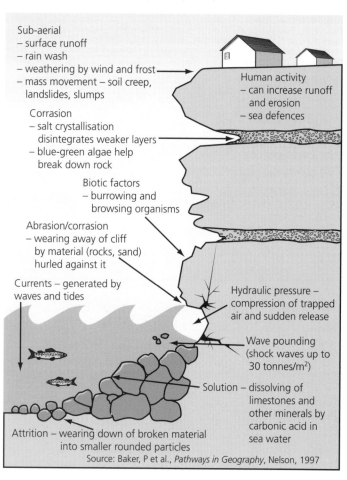

Figure 8.4 Types of erosion

Sub-aerial, or cliff-face, processes include:

- **salt weathering** – the process by which sodium and magnesium compounds expand in joints and cracks, thereby weakening rock structures
- **freeze–thaw weathering** – the process whereby water freezes, expands and degrades jointed rocks
- **biological weathering** – carried out by molluscs, sponges and urchins, and very important on low-energy coasts
- **solution weathering** – the chemical weathering of calcium by acidic water; it tends to occur in rockpools due to the presence of organisms secreting organic acids
- **slaking** – materials disintegrating when exposed to water, which can be caused by hydration cycles

Mass movements are also important in coastal areas, especially slumping and rock falls (see pages 45–49).

Wave transportation and deposition

Sediment sources are varied and include:

- onshore transport by waves
- longshore transport by waves
- rivers
- glacial and periglacial deposits
- wind-blown deposits
- artificial beach replenishment

Sediment transport is generally categorised into two modes:

- **Bedload** – grains transported by bedload are moved with continuous contact (**traction** or dragging) or by discontinuous contact (**saltation**) with the seafloor. Traction, in which grains slide or roll along, is a slow form of transport.
- **Suspended load** – grains are carried by turbulent flow and generally are held up by the water. Suspension occurs when moderate currents are transporting silts or strong currents are transporting sands.

Deposition is governed by sediment size (mass) and shape. In some cases sediments will flocculate (stick together), become heavier and fall out in deposition.

Sediment cells

The coastal sediment system, or **littoral cell system**, is a simplified model that examines coastal processes and patterns in a given area (Figure 8.5). It operates at a variety of scales from a single bay, e.g. Turtle Bay, North Queensland, Australia, to a regional scale, e.g. the south California coast. Each littoral cell is self-contained, with inputs and outputs balanced.

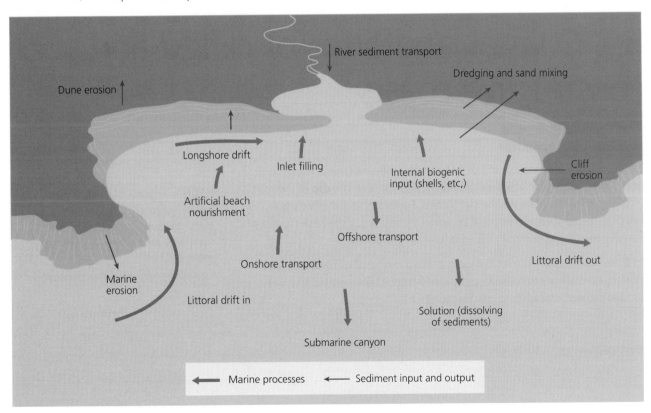

Figure 8.5 Sediment cells

If refraction is not complete, longshore drift occurs (Figure 8.6). This leads to a gradual movement of sediment along the shore, as the swash moves in the direction of the prevailing wind, whereas the backwash moves straight down the beach, following the steepest gradient.

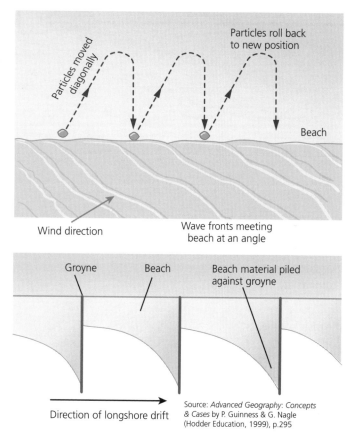

Source: *Advanced Geography: Concepts & Cases* by P. Guinness & G. Nagle (Hodder Education, 1999), p.295

Figure 8.6 Longshore drift

Now test yourself

5 Explain what is meant by the term 'littoral cell' (sediment cell).

6 Distinguish between bedload and suspended load.

Answers on p.218

Tested

8.2 Coastal landforms of cliffed and constructive coasts

Cliffs and erosion
Revised

Cliff profiles are very variable and depend on a number of controlling factors. One major factor is the influence of bedding and jointing. The **dip** of the bedding alone will create varying cliff profiles. For example, if the beds dip vertically, then a sheer cliff face will be found. By contrast, if the beds dip steeply seaward, then steep, shelving cliffs with landslips will be found.

To some extent each cliff profile is unique, but a model of cliff evolution or modification has been produced to take into account not only wave activity, but also sub-aerial weathering processes (Figure 8.7).

> A **cliff** is a rock-face along the coast, where coastal erosion, weathering and mass movements are active and the slope rises steeply (over 45°) and for some distance.

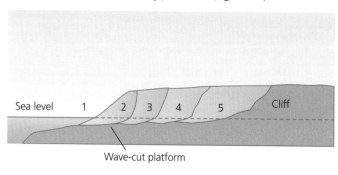

Source: Goudie, A, *The Nature of the Environment*, Blackwell, 1993

Figure 8.7 Evolution of wave-cut platforms

Many cliffs are composed of more than one rock type. These are known as composite cliffs. The exact shape and form of the cliff will depend on such factors as strength and structure of rock, relative hardness and the nature of the waves involved.

A model of coastal erosion is shown in Figure 8.8.

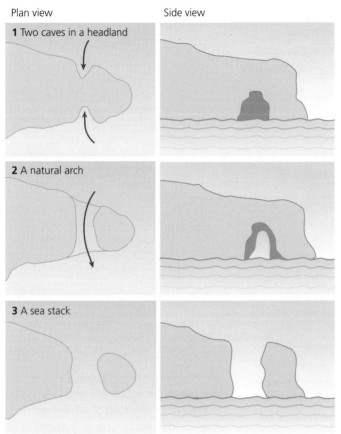

1 Wave refraction concentrates erosion on the sides of headlands. Weaknesses such as joints or cracks in the rock are exploited, forming caves.
2 Caves enlarge and are eroded further back into the headland until eventually the caves from each side meet and an arch is formed.
3 Continued erosion, weathering and mass movements enlarge the arch and cause the roof of the arch to collapse, forming a high standing stack.

Figure 8.8 Caves, arches, stacks and stumps

As a result of cliff retreat, a platform along the coast is normally created. Traditionally, this feature was described as a **wave-cut platform** (or abrasion platform), because it was believed that it was created entirely by wave action. However, there is some debate over the importance of other agents of weathering and erosion in the production of the coastal platform, especially the larger ones:

● In high latitudes, **frost action** could be important in supporting wave activity, particularly as these areas are now rising as a result of isostatic recovery (after intense glaciation).

● In other areas, **solution weathering**, **salt crystallisation** and **slaking** could also support wave activity, particularly in the tidal zone and splash zone.

● **Marine organisms**, especially algae, can accelerate weathering at low tide and in the area just above HWM. At night carbon dioxide is released by algae because photosynthesis does not occur. This carbon dioxide combines with the cool sea water to create an acidic environment, causing solution weathering.

Expert tip

Wave-cut platforms can be formed in numerous ways – wave erosion, frost action, weathering and organic action could all be responsible.

Deposition

It is important to distinguish between two types of coastline:

- **Swash-aligned coasts** are orientated parallel to the crests of the prevailing waves (Figure 8.9a). They are closed systems in terms of longshore drift transport and the net littoral drift rates are zero.
- **Drift-aligned coasts** are orientated obliquely to the crest of the prevailing waves (Figure 8.9b). The shoreline of drift-aligned coasts is primarily controlled by longshore sediment transport processes. Drift-aligned coasts are open systems in terms of longshore drift transport. Therefore, spits, bars and tombolos are all features of drift-aligned coasts.

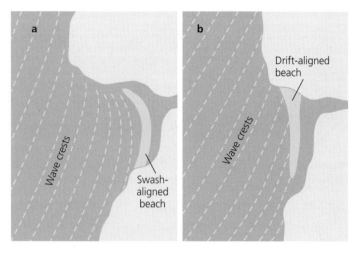

Figure 8.9 Swash-aligned (a) and drift-aligned (b) coasts

A **beach** refers to the accumulation of material deposited between the lowest tide and the highest spring tides (Figure 8.10). Beach form is affected by the size, shape and composition of materials, tidal range and wave characteristics.

As storm waves are more frequent in winter and swell waves more important in summer many beaches differ in their winter and summer profile (Figure 8.11). Thus the same beach may produce two very different profiles at different times of the year, for example constructive waves in summer may build up the beach but destructive waves in winter may break it down again.

> A **beach** is a feature of coastal deposition, consisting of pebbles on exposed coasts or sand on sheltered coasts.

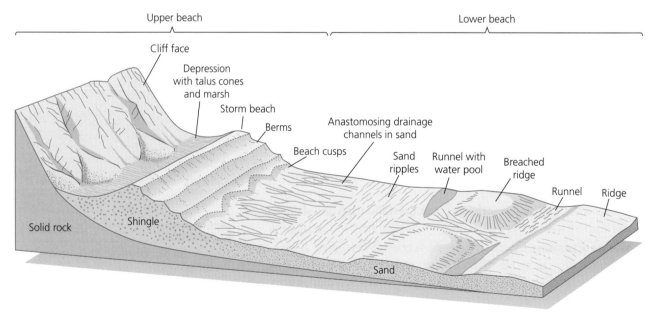

Figure 8.10 Beach deposits

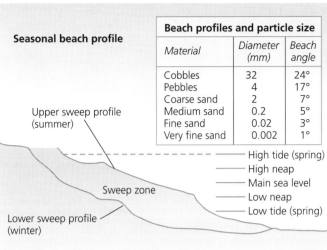

Beach profiles and particle size		
Material	*Diameter (mm)*	*Beach angle*
Cobbles	32	24°
Pebbles	4	17°
Coarse sand	2	7°
Medium sand	0.2	5°
Fine sand	0.02	3°
Very fine sand	0.002	1°

Figure 8.11 Seasonal changes on a beach

Spits and bars

These localised depositional features (Figure 8.12) will develop where:

● abundant material is available, particularly shingle and sand

● the coastline is irregular due, for example, to local geological variety (transverse coast)

● deposition is increased by the presence of vegetation (reducing wave velocity and energy)

● there are estuaries and major rivers

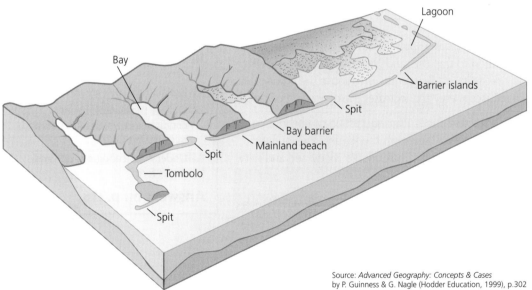

Source: *Advanced Geography: Concepts & Cases*
by P. Guinness & G. Nagle (Hodder Education, 1999), p.302

Figure 8.12 Localised depositional features

Spits are common along indented coastlines. For example, along a transverse coast, where bays are common, or near mouths (estuaries and rias), wave energy is reduced. The long, narrow ridges of sand and shingle that form spits are always joined at one end to the mainland. The simplest spit is a linear spit, which is predominantly a linear feature, but it may be curved at its distal (unattached) end.

Spits often become curved as waves undergo refraction. Cross currents or occasional storm waves may assist this hooked formation. If the curved end is very pronounced it is known as a **recurved spit**. Many spits have developed over long periods of time and have complex morphologies. An example of complex morphology is a **compound recurved spit**. This has a narrow proximal (joined) end and a wide, recurved distal end that often encloses a lagoon. The wide distal end usually consists of several dune/beach ridges associated with older shorelines, demonstrating seaward migration of the shoreline.

Spits grow in the direction of the predominant longshore drift, and are a classic example of a drift-aligned feature that can only exist through the continued supply of sediment.

Within the curve of the spit, the water is shallow and a considerable area of mudflat and saltmarsh (**salting**) is exposed at low water. These saltmarshes continue to grow as mud is trapped by the marsh vegetation. The whole area of saltmarsh is intersected by a complex network of channels (**creeks**), which contain water even at low tide.

A **bar** is a ridge of material that is connected at both ends to the mainland. It is located above sea level. If a spit continues to grow lengthwise, it may ultimately link two headlands to form a **bay bar**.

Offshore bars are usually composed of coarse sand or shingle. They develop on a gently-shelving sea-bed. Waves 'feel bottom' far offshore. This causes disturbance in the water, which leads to deposition, forming an off-shore bar below sea level. Between the bar and shore, lagoons (often called **sounds**) develop. If the lagoonal water is calm and fed by rivers, marshes and mud flats can be found. Bars can be driven onshore by storm winds and waves.

Barrier Islands are natural sandy breakwaters that form parallel to flat coastlines. By far the world's longest series is that of roughly 300 islands along the east and southern coasts of the USA. If a ridge of material links an island with the mainland, this ridge is called a **tombolo**. A classic example is Chesil Beach in Dorset.

Sand dunes

Sand dunes form where there is a reliable supply of sand, strong on-shore winds, a large tidal range and vegetation to trap the sand. Vegetation causes the wind velocity to drop, especially in the lowest few centimetres above the ground, and this reduction in velocity reduces energy and increases the deposition of sand.

Vegetation is also required to stabilise dunes. Plant succession can be interpreted by the fact that the oldest dunes are furthest from the sea and the youngest ones are closest to the shore. On the shore, conditions are windy, arid and salty.

Tidal estuaries and salt marshes

Salt marshes are typically found in three locations – along low-energy coastlines, behind spits and barrier islands, and in estuaries and harbours.

Silt accumulates in these situations and, on reaching sea-level, forms mudbanks. With the appearance of vegetation, salt marsh is formed. The mud banks are often intersected by creeks.

The intertidal zone, the zone between high tide and low tide, experiences severe environmental changes in salinity, tidal inundation and sediment composition. Halophytic (salt-tolerant) plants have adapted to the unstable, rapidly changing conditions.

8.3 Coral reefs

The development and distribution of coral

Revised

Coral reefs are calcium carbonate structures, made up of reef-building corals. Coral is limited by the depth of light penetration, so reefs occur in shallow water, up to 60 m. This dependence on light also means reefs are only found where the

---

Typical mistake

Bars can be formed as a result of longshore drift, but they can also be formed as a result of rising sea levels drowning former sand dunes and beaches.

Now test yourself

9 Describe the main features of a spit.

10 Distinguish between tombolos and bars.

11 Explain how spits are formed.

12 Outline the environmental conditions in which mud flats and salt marshes occur.

13 Under what conditions do sand dunes form?

Answers on p.218

Tested

Expert tip

This is a relatively narrow part of the syllabus, so make sure that you know it in detail.

surrounding waters contain relatively small amounts of suspended material (and are paradoxically of low productivity).

Although corals are found quite widely, reef-building corals live only in tropical seas, where temperature, salinity and lack of turbid water are conducive to their existence.

All tropical reefs began life as polyps – tiny, soft animals, like sea anemones – which attach themselves to a hard surface in shallow seas where there is sufficient light for growth. As they grow, many of these polyps exude calcium carbonate, which forms their skeleton. Then as they grow and die the reefs are built up.

Polyps have small algae, **zooxanthellae**, growing inside them.

There is a symbiotic relationship between the polyps and the algae – i.e. both benefit form the association:

- The algae get shelter and food from the polyp, while the polyp also gets some food via photosynthesis (turning light energy from the sun into food).
- This photosynthesis means that algae need sunlight to live, hence corals only grow where the sea is shallow and clear.

The distribution of coral (Figure 8.13) is controlled by seven main factors:

- **Temperature** – no reefs develop where the mean annual temperature is below 20°C. Optimal conditions for growth are between 23°C and 25°C.
- **Depth of water** – most reefs grow in depths of water less than 25 m, and so are generally found on the margins of continents and islands.
- **Light** – corals prefer shallow water because they need light for the zooxanthellae, which supply the coral with as much as 98% of its food requirements.
- **Salinity** – corals are marine organisms and are intolerant of water with salinity levels below 32 psu, but can tolerate high salinity levels (>42 psu) as found in the Red Sea or the Persian Gulf.
- **Sediment** – sediment has a negative effect on coral. It clogs up feeding structures and cleansing systems and sediment-rich water reduces the light available for photosynthesis.
- **Wave action** – coral reefs generally prefer strong wave action, which ensures oxygenated water and a stronger cleansing action. This helps remove any trapped sediment and also supplies microscopic plankton to the coral. However, in extreme conditions, south as the South Asian tsunami, the waves may be too destructive for the coral to survive.
- **Exposure to the air** – corals die if they are exposed to the air for too long. They are therefore mostly found below the low tide mark.

> **Coral** is a living organism (polyp) found in clear, tropical waters. Corals live in communities known as reefs. They secrete lime and build a skeleton.

Typical mistake

> Many students think that coral is a rock – it is a living organism.

> **Zooxanthellae** are algae that live in a symbiotic relationship with coral and give it its colour.

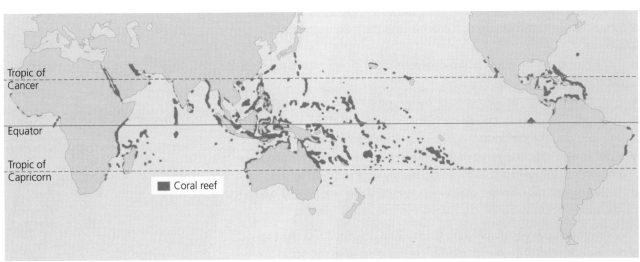

Figure 8.13 World distribution of coral reefs

Types of coral

Fringing reefs lie along the coast of a landmass. They are usually characterised by an outer reef edge capped by an algal ridge, a broad reef flat and a sand-floored 'boat channel' close to the shore. Many fringing reefs grow along shores that are protected by barrier reefs and are thus characterised by organisms that are best adapted to low-wave-energy conditions.

Barrier reefs occur at greater distances from the shore than fringing reefs and are commonly separated from it by a wide, deep lagoon. Barrier reefs tend to be broader, older, and more continuous than fringing reefs; the largest barrier-reef system in the world is the Great Barrier Reef, which extends 1600 km along the east Australian coast, usually tens of kilometres offshore.

Atolls rise from submerged volcanic foundations and often support small islands of wave-borne detritus. Atolls are essentially indistinguishable in form and species composition from barrier reefs except that they are confined to the flanks of submerged oceanic islands, whereas barrier reefs may also flank continents.

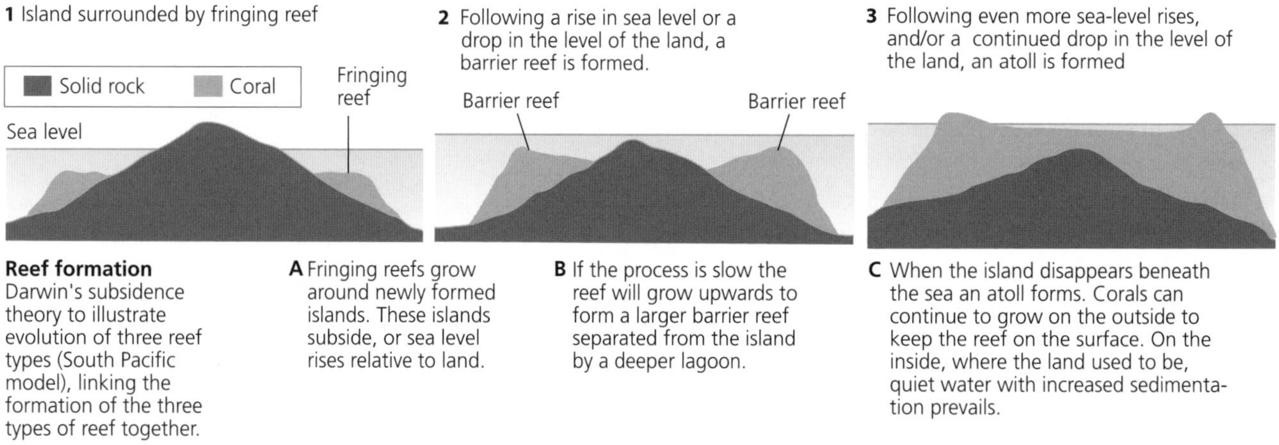

1 Island surrounded by fringing reef

2 Following a rise in sea level or a drop in the level of the land, a barrier reef is formed.

3 Following even more sea-level rises, and/or a continued drop in the level of the land, an atoll is formed

Reef formation
Darwin's subsidence theory to illustrate evolution of three reef types (South Pacific model), linking the formation of the three types of reef together.

A Fringing reefs grow around newly formed islands. These islands subside, or sea level rises relative to land.

B If the process is slow the reef will grow upwards to form a larger barrier reef separated from the island by a deeper lagoon.

C When the island disappears beneath the sea an atoll forms. Corals can continue to grow on the outside to keep the reef on the surface. On the inside, where the land used to be, quiet water with increased sedimentation prevails.

Figure 8.14 Formation of fringing reefs, barrier reefs and atolls

Origins

The origin of fringing reefs is quite clear – they simply grow seaward from the land. Barrier reefs and atolls, however, seem to rise from considerable depth, far below the level at which coral can grow, and many atolls are isolated in deep water. The lagoons between a barrier reef and the coast, usually 45–100 m in depth, and often many kilometres wide, require some explanation.

In 1842 Charles Darwin, supported by Dana and others, explained the growth of barrier reefs and atolls as a gradual process, with the main reason being subsidence:

- Darwin outlined the way in which coral reefs could grow upwards from submerging foundations.
- Fringing reefs might be succeeded by barrier reefs and thence by atolls.
- A fringing reef grows around an island, for example, and as this slowly subsides, the coral continues to grow, keeping pace with the subsidence.
- Coral growth is more vigorous on the outer side of the reef so it forms a higher rim, whereas the inner part forms an increasingly wide and deep lagoon.
- Eventually the inner island is submerged, forming a ring of coral reef, which is the atoll.

An alternative theory was that of John Murray, who suggested that the base of the reef consisted of a submarine hill or plateau rising from the ocean floor. These reached within 60 m of the of the sea surface and consisted of either sub-surface volcanic peaks or wave-worn stumps.

Reginald Daly suggested that a rise in sea level might be responsible:

● A rise did take place in post-glacial times as ice sheets melted.
● The sea would have been much colder and much lower (about 100 m) during glacial times.
● All coral would have died, and any coral surfaces would have been eroded down by the sea.
● Once conditions started to warm, and sea level was rising, the previous coral reefs provided a base for the upward growth of coral.

Darwin's theory still receives considerable support. While Daly was correct in principle, it is now believed that the erosion of the old reefs was much less rapid than previously believed.

8.4 Sustainable development of coasts

The Soufrière Marine Management Area, St Lucia
Revised

The coastline around Soufrière, St Lucia is very popular with tourists, yachtsmen, fishermen, divers and local people. Increasingly, as tourism and fishing have become the two most important economic sectors in the region, there has been an increase in the number of conflicts between the users for the limited space and resources. For example:

● yachtsmen and fishermen competed for the use of marine space for mooring and netting activities, respectively
● divers were often accused by fishermen of deliberately damaging fish pots/traps found during dive expeditions and disrupting coral reefs
● researchers were accused by fishermen of taking fish and coral reef samples, thus contributing to environmental degradation
● local residents had conflicts with local hoteliers over access to beach areas for fishing and recreation
● fishermen had conflicts with the tourism sector and management authorities over the location of a jetty in the Soufrière Bay to facilitate tourism-related traffic; the jetty was seen as an obstruction to their nets
● tourism-related vessel operators (water taxis and one-day cruisers) were accused by fishermen of interrupting fishing and damaging nets and traps
● there were problems of visitor harassment by disorganised water taxi operators offering their services
● there was evidence of indiscriminate anchorage on coral reefs by yachtsmen
● entry into fragile habitat areas by divers was unregulated
● the decrease in nearshore fisheries and resources in general was becoming increasingly apparent
● uncoordinated and unauthorised marine scientific research was reported
● degradation of coastal water quality was a problem, with direct ramifications for human health and the integrity of marine ecosystems
● degradation of coastal landscapes was becoming apparent
● solid waste accumulation (especially plastics) was a cause for concern
● there was a general lack of awareness of, and appreciation for, the marine environment

To manage these issues the Soufrière Marine Management Area was established in 1986. Since then its aims and objectives have evolved. The SMMA mission statement is as follows:

The mission of the Soufrière Marine Management Area is to contribute to national and local development, particularly in the fisheries and tourism sectors through

Sustainable development increases the standard of living of the population but does not compromise the needs of future generations.

Expert tip

When learning about your chosen case study, practise a sketch map of the area but also find out some background information. For example what is the population size? What is the growth rate? What are the main forms of economic activity?

management of the Soufrière coastal zone based on the principles of sustainable use, cooperation among resource users, institutional collaboration, active and enlightened participation, and equitable sharing of benefits and responsibilities among stakeholders.

The SMMA agreement has established five different types of zone within the area:

1 Marine reserves: these are areas for the purpose of protecting the natural resources they contain. No extractive activity is allowed and entry into a reserve is subject to approval by the Department of Fisheries.

2 Fishing priority areas: these are areas maintaining and sustaining fishing activities, which take priority over any other use of the area.

3 Recreational areas: these are beaches and marine environments (for swimming and snorkelling), which are reserved for public access and recreation.

4 Yachting areas: specific areas are designated to facilitate the use of pleasure boats and yachts and for the protection of the bottom substrate.

5 Multiple use areas: these areas are where fishing, diving, snorkelling and other recreational activities can take place.

This **land use zoning** system (Figure 8.15) caters for the array of users of the area and yet provides protection for some of the island's critical marine resources.

> **Typical mistake**
>
> The SMAA is an ongoing scheme – conditions may change over time. If it is successful, it may attract more people to the area, and so increase local pressures there.

> **Land use zoning** is the allocation of different sections of land or sea for different, often conflicting, users.

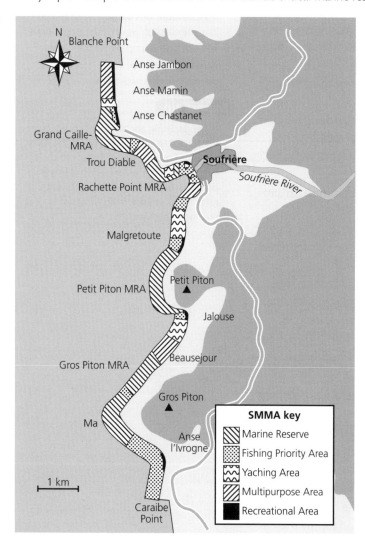

Figure 8.15 Land use zoning in the SMMA

The scheme has been successful. Fish stocks have increased; the balance between tourism and fisheries has been achieved; and there has been an improvement in the quality of coral in the area. Waste disposal has also been improved.

Now test yourself

Tested

16 Define sustainable development.

17 Identify two conflicting users of the Soufrière Marine Management Area.

18 How has the SMMA been managed?

Answers on p.218

Exam-style questions

1 **(a)** Outline the main characteristics of constructive and destructive waves.

  (b) Describe and explain variations in cliff profiles.

2 **(a)** Describe and explain the formation of different types of coral reef.

  (b) To what extent is it possible to manage coastal areas sustainably?

Exam ready

9 Hazardous environments

9.1 Hazardous environments resulting from crustal (tectonic) movement

Global distribution of tectonic hazards
Revised

Tectonic hazards include seismic activity (earthquakes), volcanoes and tsunamis. Most of the world's earthquakes occur in clearly defined linear patterns (Figure 9.1). These linear chains generally follow plate boundaries.

Earthquakes

Broad belts of earthquakes are associated with subduction zones (where a dense ocean plate plunges beneath a less dense continental plate) whereas narrower belts of earthquakes are associated with constructive plate margins, where new material is formed, and plates are moving apart.

Collision boundaries, such as in the Himalayas, are also associated with broad belts of earthquakes, whereas conservative plate boundaries, such as California's San Andreas fault line, give a relatively narrow belt of earthquakes (this can still be over 100 km wide). In addition, there appear to be occurrences of earthquakes related to isolated plumes of tectonic activity, known as hotspots.

> **Typical mistake**
>
> Although most earthquakes are associated with plate boundaries and tectonic activity, many earthquakes occur at great distances from plate boundaries and are not readily explained by tectonic activity.

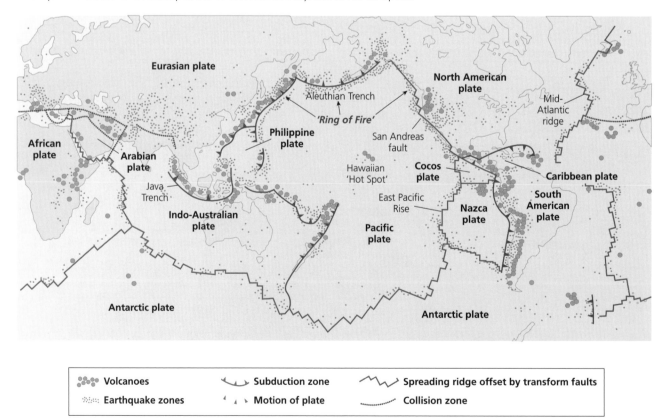

Figure 9.1 Distribution of plates, plate boundaries, volcanoes and earthquakes

Volcanoes

Most volcanoes are found at plate boundaries although there are some exceptions, such as the volcanoes of Hawaii, which occur over hot spots (isolated plumes of rising magma). About three-quarters of the Earth's 550 historically active volcanoes lie along the Pacific Ring of Fire.

At subduction zones volcanoes produce more viscous lava, and tend to erupt explosively and produce much ash. By contrast, volcanoes that are found at mid-ocean ridges or hot spots tend to produce relatively fluid basaltic lava, as in the case of Iceland and Hawaii.

Tsunamis

Up to 90% of the world's tsunamis occur in the Pacific Ocean. This is because they are associated with subduction zones, most of which are found in the Pacific.

Tsunamis are generally caused by earthquakes (usually in subduction zones) but can be caused by volcanoes (e.g. Krakatoa, 1883) and landslides (e.g Alaska, 1964).

Earthquakes and resultant hazards

Revised

Table 9.1 Earthquake hazards and impacts

Hazards	Impacts
Primary hazards: ● Ground shaking ● Surface faulting Secondary hazards: ● Ground failure and soil liquefaction ● Landslides and rockfalls ● Debris flow and mud flow ● Tsunamis	Loss of life Loss of livelihood Total or partial destruction of building structure Interruption of water supplies Breakage of sewage disposal systems Loss of public utilities such as electricity or gas Floods from collapsed dams Release of hazardous material Fires Spread of chronic illness

Factors affecting earthquake damage

The extent of earthquake damage is influenced by:

● the strength and depth of earthquake and number of aftershocks

● population density

● the type of buildings

● the time of day

● the distance from the centre (epicentre) of the earthquake

● the type of rocks and sediments

● secondary hazards

● economic development

Most earthquakes occur with little, if any, advance warning. Most problems are associated with damage to buildings, structures and transport.

Dealing with earthquakes

The main ways of dealing with earthquakes involve better forecasting techniques, warning systems and emergency procedures, and improvements to building design and location.

There are a number of ways of predicting and monitoring earthquakes. These include measurement of:

- small-scale uplift, subsidence or ground tilt
- changes in rock stress
- microearthquake activity (clusters of small quakes)
- anomalies in the Earth's magnetic field
- changes in radon gas concentration
- changes in electrical resistivity of rocks

Tsunami warning systems

At present it is impossible to predict precisely where and when a tsunami will happen. In most cases it is only possible to raise the alarm once a tsunami has started (early warning system).

Now test yourself

1 What are the main hazards associated with earthquake activity?
2 In what ways is it possible to predict and monitor earthquakes?
3 Outline the causes of tsunamis.
4 To what extent is it possible to manage the impacts of tsunamis?

Answers on pp.218–219

Tested

Volcanic hazards

Revised

Volcanic hazards can be divided into six main categories

- lava flows
- ballistics and tephra clouds
- pyroclastic flows and ash fallout
- gases and acid rain
- lahars (mud flows)
- glacier bursts (jokulhlaups)

A **pyroclastic flow** is a fast-moving cloud of extremely hot gas, ash and rock fragments, which can reach temperatures of about 1000°C and travel at speeds of up to 700 km/h.

Table 9.2 Hazards associated with volcanic activity

Direct hazards (primary hazards)	Indirect hazards (secondary hazards)	Socio-economic impacts
Pyroclastic flows	Atmospheric ash fall out	Destruction of settlements
Volcanic bombs (projectiles)	Landslides	Loss of life
Lava flows	Tsunamis	Loss of farmland and forests
Ash fallout	Acid rainfall	
Volcanic gases	Lahars (mudflows)	Destruction of infrastructure – roads, airstrips and port facilities
Nuees ardentes		
Earthquakes		Disruption of communications

A **nuee ardente** is a mass of hot gas, superheated steam and volcanic dust that travels down the side of a volcano as a 'glowing avalanche' following a volcanic eruption.

Expert tip

It is a good idea to have contrasting case studies of volcanoes – one MEDC and one LEDC – to bring out differences in impact, management, land-use etc.

Predicting volcanoes

Volcanoes are easier to predict than earthquakes because there are certain signs. The main ways of predicting volcanoes include:

- seismometers to record swarms of tiny earthquakes that occur as the magma rises
- chemical sensors to measure increased sulfur levels
- lasers to detect the physical swelling of the volcano
- ultrasound to monitor low-frequency waves in the magma, resulting from the surge of gas and molten rock
- direct observation

Now test yourself

5 What are the main hazards associated with volcanoes?
6 In what ways is it possible to predict volcanoes?

Answers on p.219

Tested

Factors affecting the perception of risk

At an individual level there are three important influences on an individual's response to risk. For example:

- experience – the more experience of environmental hazards the greater the adjustment to the hazard
- material well-being – those who are better off have more choice
- personality – is the person a leader or a follower, a risk-taker or risk-minimiser?

Ultimately there are three choices: do nothing and accept the hazard; adjust to the situation of living in a hazardous environment; leave the area. It is the adjustment to the hazard that we are interested in. The level of adjustment will depend, in part, upon assessing the risks caused by the hazard. This includes:

- identification of the hazards
- estimation of the risk (probability) of the environmental hazard
- evaluation of the cost (loss) caused by the environmental hazard

> **Now test yourself**
>
> 7 What are the factors that influence an individual's response to risk?
> 8 What choices do people have with regard to living in areas at risk of natural hazards?
>
> **Answers on p.219**
>
> Tested

9.2 Hazardous environments resulting from mass movements

Causes of mass movements

Mass movements are a common natural event in unstable, steep areas. They can lead to loss of life, disruption of transport and communications, and damage to property and infrastructure. The most important factors that determine movement are gravity, slope angle and pore pressure. Increases in shear stress and/or decreases in shear resistance trigger mass movement (Table 9.3).

> **Mass movement** is any large-scale movement of the Earth's surface that is not accompanied by a moving agent.

Table 9.3 Factors contributing to increasing shear stress and decreasing shear resistance

Factors contributing to increased shear stress	Examples
Removal of lateral support through undercutting or slope steepening	Erosion by rivers and glaciers, wave action, faulting, previous rock falls or slides
Removal of underlying support	Undercutting by rivers and waves, subsurface solution, loss of strength by exposure of sediments
Loading of slope	Weight of water, vegetation, accumulation of debris
Lateral pressure	Water in cracks, freezing in cracks, swelling, pressure release
Transient stresses	Earthquakes, movement of trees in wind
Factors contributing to reduced shear strength	**Examples**
Weathering effects	Disintegration of granular rocks, hydration of clay minerals, solution of cementing minerals in rock or soil
Changes in pore-water pressure	Saturation, softening of material
Changes of structure	Creation of fissures in clays, remoulding of sands and clays
Organic effects	Burrowing of animals, decay of roots

Human activities can increase the risk of mass movements, for example by:

- increasing the slope angle by cutting through high ground – slope instability increases with slope angle
- placing extra weight on a slope (e.g. new buildings); this adds to the stress on a slope

> **Expert tip**
>
> Shear stress refers to the forces trying to pull a mass downslope, while shear resistance is the internal resistance of a slope.

- removing vegetation – roots bind the soil together and interception by leaves reduces rainfall compaction
- exposing rock joints and bedding planes, which can increase the speed of weathering

There have been various attempts to manage the hazard of mass movements. Methods to combat mass movements are largely labour intensive and include:

- building restraining structures such as walls, piles, buttresses and gabions – these can hold back minor landslides
- excavating and filling steep slopes to produce gentler ones – this can reduce the impact of gravity on a slope
- draining slopes to reduce the build-up of water – this decreases pore-water pressure in the soil
- watershed management, for example afforestation and agroforestry ('farming the forest') – this increases interception and reduces overland flow

---

Now test yourself

9 How can human activity increase the risk of mass movements?

Answer on p.219

Tested

Case study   **The Italian mudslides of 1998**

In May 1998 mudslides swept through towns and villages in Campania, Italy killing nearly 300 people. Worst affected was Sarno, a town of 35,000 people. Up to a year's rainfall had fallen in the two preceeding weeks.

Campania is Italy's most vulnerable region. Since 1892, scientists have recorded over 1170 serious landslides in Campania and Calabria. Geologically the area is unstable. It has active volcanoes, such as Vesuvius, many mountains and scores of fast-flowing rivers.

The disaster was only partially natural; much of it was down to human error:
- The River Sarno's bed had been cemented over.

- The clay soils of the surrounding mountains had been rendered dangerously loose by forest fires and deforestation.
- Houses had been built on hillsides identified as landslide zones.
- Over 20% of the houses in Sarno were built without permission.
- Most were built over a 2-metre-thick layer of lava formed by the eruption of Vesuvius in 79 AD. Heavy rain can make it liquid and up to 900 million tonnes of land are washed away in this way every year.

Hence, much of the region's fragility is due to mass construction, poor infrastructure and poor planning.

It is likely that the landslides of northern Italy will be mirrored by landslides in the Mediterranean region as it becomes more developed. All across southern Europe human impacts have combined to increase the mass movement hazard:

- The first step involves clearing the land for development. The easiest way for this is a forest fire. Large numbers of fires are started deliberately by developers to ensure that the areas that they target lose their natural beauty. One of the side effects of the fire is to loosen the underlying soil.
- In Sicily up to 20,000 holiday homes have been built on beaches, cliffs and wetlands in defiance of planning regulations.
- In Italy over 200,000 houses have been built without permission, and many are without proper drainage or foundations.
- Many stand close to riverbeds that seem empty and remain empty until storms occur.

Now test yourself

10 What are the natural reasons why northern Italy is at risk from mudslides?

11 What human factors have increased the risk of mudslides in the region?

Answers on p.219

Tested

Avalanches

Revised

Avalanches are mass movements of snow and ice. Average speeds are 40–60 km/h, but speeds of up to 200 km/h have been recorded in Japan.

Loose avalanches, comprising fresh snow, usually occur soon after a snowfall. By contrast, slab avalanches occur at a later date, when the snow has developed some cohesion. They are usually much larger than loose avalanches and cause more destruction. They are often started by a sudden rise in temperature, which causes melting. This lubricates the slab and makes it unstable.

Many of the avalanches occur in spring when the snowpack is large and temperatures are rising. There is also a relationship between the number of avalanches and altitude. For example in the Swiss Alps most occur between 2000 m and 2500 m and there is reduced occurrence both higher up and lower down.

Although avalanches cannot be prevented, it is possible to reduce their impact – various methods are shown in Figure 9.2.

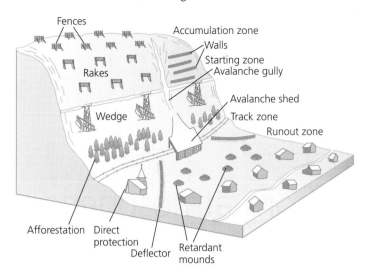

Figure 9.2 Avalanche protection schemes

Case study | **The European avalanches of 1999**

The avalanches that killed 18 people in the Alps in February 1999 were the worst in the area for nearly 100 years. Moreover, they occurred in an area that was thought to be fairly safe. In addition, precautionary measures had been taken, such as an enormous avalanche wall to defend the village of Taconnaz. However, the villages of Montroc and Le Tour, located at the head of the Chamonix Valley, had no such defences.

The avalanche that swept through the Chamonix Valley killed 11 people and destroyed 18 chalets (Figure 9.3). It was about 150 m wide, 6 m high and travelled at a speed of up to 90 km/h. Rescue work was hampered by the low temperatures (–7°C), which caused the snow to compact, and made digging almost impossible.

Nothing could have been done to prevent the avalanche and avalanche warnings had been given the day before, as the region had experienced up to 2 m of snow in just 3 days. Ongoing avalanche monitoring meant that villagers and tourists in the 'safe' zone thought that they were safe.

Buildings in Montroc were classified as being almost completely free of danger. By contrast, in the avalanche danger zones no new buildings had been developed for many decades.

Meteorologists have suggested that disruption of weather patterns resulting from global warming will lead to increased snow falls in the Alps, which will be heavier and later in the season. This would mean that the conventional wisdom regarding avalanche safe zones would need to be re-evaluated.

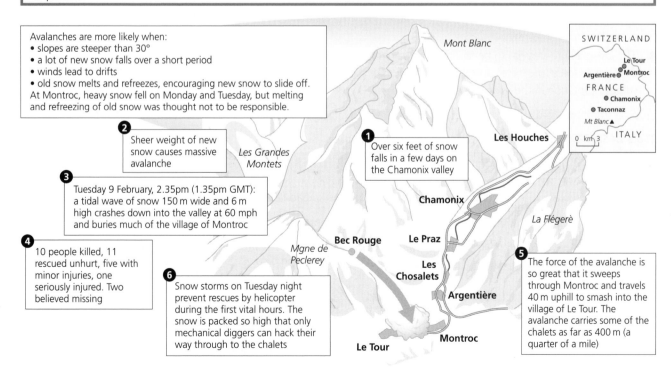

Figure 9.3 Causes and consequences of the Montroc/Le Tour avalanche

Now test yourself [Tested]

12 Suggest reasons why avalanches are clustered in the months between January and March. Explain at least **two** of these reasons.

13 What is an avalanche?

14 What are the factors that increase the risk of an avalanche?

15 What were the conditions in Europe in February 1999 that led to widespread avalanches?

16 How and why may the threat of avalanches change in the next decades?

Answers on p.219

9.3 Hazards resulting from atmospheric disturbances

Tropical storms (cyclones) [Revised]

Tropical storms bring intense rainfall and very high winds, which may in turn cause storm surges and coastal flooding, and other hazards such as flooding and mudslides.

They develop as intense low-pressure systems over tropical oceans. Winds spiral rapidly around a calm central area known as the eye. The diameter of the whole tropical storm may be as much as 800 km, although the very strong winds that cause most of the damage are found in a narrower belt up to 300 km wide.

In a mature tropical storm pressure may fall to as low as 880 millibars. This, and the strong contrast in pressure between the eye and outer part of the tropical storm, lead to very strong winds of over 118 km/h.

For tropical storms to form a number of conditions are needed:

- Sea temperatures must be over 27°C to a depth of 60 m (warm water gives off large quantities of heat when it is condensed – this is the heat which drives the tropical storm).
- The low-pressure area has to be far enough away from the equator so that the Coriolis force (the force caused by the rotation of the Earth) creates sufficient rotation in the rising air mass.

Tropical storms are measured using the Saffir-Simpson scale (Table 9.4).

> **Tropical storm** is a generic term that includes hurricanes (North Atlantic), cyclones (Indian Ocean and the Bay of Bengal) and typhoons (Japan). A tropical storm is a low-pressure system up to 600 km in diameter with wind speeds of up to 300 km/h (typically 160 km/h) and bringing up to 30–50 cm of rainfall.

> **Expert tip**
>
> There is no need to learn all of the details of the Saffir-Simpson scale, but it is good to know about the lower, middle and upper ends of the scale, as shown in Table 9.4.

Table 9.4 Saffir-Simpson scale of tropical storm strength and impacts

Category	Description
Category 1: Winds 119–153 km/h; storm surge generally 1.2–1.5 m above normal	No real damage to building structures. Damage primarily to unanchored mobile homes. Also, some coastal road flooding and minor pier damage.
Category 3: Winds 178–209 km/h; storm surge generally 2.7–3.6 m above normal	Some structural damage to small residences and utility buildings. Mobile homes are destroyed. Flooding near the coast destroys smaller structures with larger structures damaged by floating debris. Land below 1.5 m above mean sea level may be flooded inland 13 km or more. Evacuation of low-lying residences close to the shoreline may be necessary.
Category 5: Winds greater than 249 km/h; storm surge generally greater than 5.5 m above normal	Complete roof failure on many residences and industrial buildings. Some complete building failures with small utility buildings blown over or blown away. Complete destruction of mobile homes. Severe and extensive window and door damage. Low-lying escape routes are cut by rising water 3–5 hours before arrival of the centre of the tropical storm. Major damage to lower floors of all structures located less than 4.5 m above sea level and within 500 m of the shoreline. Massive evacuation of residential areas on low ground within 8–16 km of the shoreline may be required.

Tropical storm management

The unpredictability of tropical storm paths makes the effective management of tropical storms difficult, while the strongest storms do not always cause the greatest damage.

Tracking tropical storms

Information regarding tropical storms is received from a number of sources, including:

- satellite images
- aircraft that fly into the eye of the tropical storm to record weather information
- weather stations at ground level
- radars that monitor areas of intense rainfall

Preparing for tropical storms

There are a number of ways in which national governments and agencies can help prepare for a tropical storm. These include risk assessment, land-use control (including floodplain management) and reducing the vulnerability of structures and organisations.

Risk assessment

- Living in coastal areas increases the risk associated with tropical storms.
- The evaluation of risks of tropical storms can be shown in a hazard map. These can be based upon:
  - analysis of climatological records to determine how often cyclones have struck, their intensity and locations
  - the history of winds speeds and frequencies, heights and locations of flooding and storm surges over a period of about 50–100 years

Land use zoning

The aim is to control land use so that the most important facilities are placed in the least vulnerable areas, including floodplains.

Reducing vulnerability of structures and infrastructures

- New buildings should be designed to be wind and water resistant.
- Communication and utility lines should be located away from the coastal area or installed underground.
- Improvement of building sites includes raising the ground level to protect against flooding and storm surges.
- Protective river embankments, levées and coastal dikes should be regularly inspected for breaches due to erosion.
- Improved vegetation cover helps to reduce the impact of soil erosion and landslides, and facilitates the absorption of rainfall to reduce flooding.

In addition, there are many things that individuals can do to prepare for a tropical storm in terms of how to act during and after the event (Figure 9.4).

Now test yourself

17 Why is managing tropical storms difficult?

Answer on p.219

Tested

Before a tropical storm
- Know where your emergency shelters are.
- Have disaster supplies on hand.
- Protect your windows.
- Permanent shutters are the best protection. A lower-cost approach is to put up plywood panels.
- Trim back branches from trees.
- Trim branches away from your home and cut out all dead or weak branches on any trees on your property.
- Check your home and car insurance.
- Make arrangements for pets and livestock.
- Develop an emergency communication plan.

During a tropical storm
- Listen to the radio or television for tropical storm progress reports.
- Check emergency supplies.
- Make sure your car is full of fuel.
- Bring in outdoor objects such as lawn furniture, toys, garden tools, and anchor objects that cannot be brought inside.
- Secure buildings by closing and boarding up windows.
- Remove outside antennas and satellite dishes.

After a tropical storm
- Assist in search and rescue.
- Seek medical attention for persons injured.
- Clean up debris and effect temporary repairs.
- Report damage to utilities.
- Watch out for secondary hazards: fire, flooding, etc.

Figure 9.4 What to do before, during and after a tropical storm

Tornadoes

Revised

Tornadoes are small and short-lived but highly destructive storms. Because of their severe nature and small size, comparatively little is known about them.

Tornados consist of elongated funnels of cloud that descend from the base of a well-developed cumulonimbus cloud, eventually making contact with the ground beneath.

Figure 9.5 shows how tornados form.

> **Typical mistake**
>
> Some students think that hurricanes and tornadoes are the same – they are very different in origin, scale, impact and potential for management.

> A **tornado** is a violent, destructive weather system, with powerful rotating winds (up to 300 km/h). Tornadoes are intense, low-pressure systems, and their development depends on instability in the atmosphere, convergence and strong updrafts in the air.

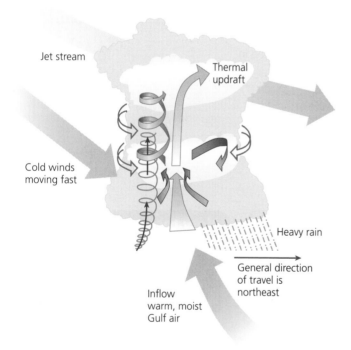

Jet stream

Thermal updraft

Cold winds moving fast

Heavy rain

General direction of travel is northeast

Inflow warm, moist Gulf air

Figure 9.5 Formation of tornadoes in the USA

Many tornadoes have a short life. They can last from several seconds to more than an hour.

'Favoured' areas are temperate continental interiors in spring and early summer, when insolation is strong and the air may be unstable, although many parts of the world can be affected by tornado outbreaks at some time or another. The Great Plains of the USA, including Oklahoma, Texas and Kansas, have a high global frequency, and are particularly prone at times when cool, dry air from the Rockies overlies warm, moist 'Gulf' air.

Tornado damage

About 1000 tornadoes hit the USA each year. On average, tornadoes kill about 60 people per year – mostly resulting from flying or falling (crushing) debris. There are three damaging factors at work:

- The winds are often so strong that objects in the tornado's path are simply removed or very severely damaged.
- Strong rotational movement tends to twist objects from their fixings, and powerful uplift can carry some debris upwards into the cloud.
- The very low atmospheric pressure near the vortex centre is a major source of damage. When a tornado approaches a building, external pressure is rapidly reduced, and unless there is a nearly simultaneous and equivalent decrease in internal pressure, the walls and roof may explode outwards in the process of equalising the pressure differences.

The Fujita scale relates the degree of damage to the intensity of the wind (Table 9.5). It should be used with caution as it does not take into account differences in building structure and materials. A new enhanced F-scale, introduced in 2006, classifies damage F0–F5 according to different types of buildings/materials.

Table 9.5 Fujita tornado damage scale

Category	Description
Category F0: Light damage (<117 km/h)	Some damage to chimneys; branches broken off trees; shallow-rooted trees pushed over; sign boards damaged
Category F3: Severe damage (254–332 km/h)	Roofs and some walls torn off well-constructed houses; trains overturned; most trees in forests uprooted; heavy cars lifted off ground and thrown
Category F5: Incredible damage (419–512 km/h)	Strong-frame houses lifted off foundations and swept away; automobile-sized missiles fly through the air in excess of 100 m; trees debarked

Now test yourself

18 To what extent is it possible to manage the risk of tornado damage?

Answer on p.219

Tested

Managing tornados

As yet there is no effective way of managing tornadoes. The best advice is to stay indoors and, if possible, underground.

There is no proof that cloud seeding can or cannot change tornado potential in a thunderstorm.

9.4 Sustainable management in hazardous environments

The use of geo-materials for erosion and sediment control, Fraser's Hill, Pahang, Malaysia

Revised

In Malaysia, research on **bioengineering** has involved studies on plant selection for the re-vegetation of cut slopes along highways. Bioengineering designs have great potential and application in Malaysia because in deforested upland sites, landslides are common, particularly during the wetter months between November and January. Fraser's Hill is an area of lower montane forest and receives 200–410 mm of rainfall each month Post-landslide restoration works involving conventional civil designs are costly and sometimes not practical at remote sites. Because of these constraints and the low risk to lives and property, bioengineering was the option taken for erosion control, slope stabilisation and vegetation establishment.

Two study plots were chosen and one control plot. Initial work involved soil nailing, using 300 live stakes of *angsana* tree branches and 200 cut stems of *ubi kayu*. Subsequently, major ground works involved the installation of geo-structures. Tall saplings of *Toona sinensis*, a fast growing tree species, were then planted at the toe of the slope for long-term stability.

Live stakes and cut stems
- At the end of 6 months, the live stakes became living trees.
- A high percentage of *angsana* stakes (93%) sprouted shoots and roots after a month, and 75% of *ubi kayu* stems sprouted leaves within a week.
- Vegetation cover on slopes helped reduce soil erosion because shoots lowered the intensity of raindrops falling on the exposed soil.
- Furthermore, roots functioned like mini soil nails to increase the shear strength of surface soils.
- Thus, live stakes were effective in stabilising unstable slopes, and their use in bioengineering should be promoted in the wet tropics.

Slope stability
- Without the erosion control measures, there was aggressive soil erosion during heavy downpours, which caused scouring of the steep slope below the tarred road and resulted in an overhang of the road shoulder.

Trapped sediments and vegetation establishment
- After one year about 75% of one study site was covered by vegetation, while 90% of the second plot was revegetated.
- There was no more incidence of landslide at these two plots.
- However, at the control plot, there was further soil erosion, which resulted in further undercutting of the slope face.
- At the control plot, after 1 year, only seven plant species were present. These were weeds.
- The poor vegetation cover is probably due to unstable soil conditions caused by frequent soil erosion and minor landslides.

Evaluation
- The **geo-structures** were installed at a cost of US$3078, which was cheaper than restoration works using conventional civil structures such as rock gabions, which would cost about US$20,000.
- As the site is fairly remote, higher transportation and labour costs would have contributed to the higher cost of constructing rock gabions at this site.
- On the other hand, the geo-materials, which were abundantly available locally, were relatively cheap to make or purchase, and this contributed to the low project cost. The geo-structures were non-polluting, required minimal post-installation maintenance, were visually attractive and could support greater biodiversity within the restored habitats.
- The **geo-materials** used in the project, such as coir rolls and straw wattles, biodegrade after about a year and become organic fertilisers for the newly established vegetation.
- After 18 months, the restored cut slopes were almost covered by vegetation, and there was no further incident of landslides.

Bioengineering is the use of vegetation in engineering, for example the selection of suitable plant species for the recolonisation of areas following landslides.

Geo-materials refer to the use of naturally occurring materials such as vegetation in engineering.

Geo-structures are structures constructed from geo-materials such as bamboo bundles (fachines), coir rolls and straw wattles.

Now test yourself

Tested ☐

19 Evaluate the success of one management scheme in a hazardous environment that you have studied.

Answer on p.219

Expert tip

Your case study should be named *and* located.

Exam-style questions

1 **(a)** Outline the range of hazards associated with volcanic eruptions. [10]

 (b) How does the impact of hurricanes vary with levels of economic development? [15]

2 **(a)** Outline the natural and human reasons why mass movements occur. [10]

 (b) Describe ways in which it is possible to limit mass movements. [15]

Exam ready ☐

Typical mistake

Some students forget to comment about the control study – to evaluate the success of a research project there needs to be a contrast between the experimental and control studies.

10 Arid and semi-arid environments

10.1 The distribution and climatic characteristics of hot arid and semi-arid environments

Aridity

Revised

While Africa contains the greatest proportion of the global arid zone (Figure 10.1), Australia is the most arid continent with about 75% of the land being classified as **arid** or **semi-arid**.

Most arid areas are located in the tropics, associated with the sub-tropical high pressure belt. However, some are located alongside cold ocean currents (such as the Namib and Atacama deserts), some are located in the lee of mountain ranges (such as the Gobi and Patagonian deserts) while others are located in continental interiors (such as the Sahara and the Australian deserts).

> **Arid** refers to areas with a permanent water deficit and less than 250 mm of rainfall per annum.
>
> **Semi-arid** is commonly defined as having rainfall of less than 500 mm per annum.

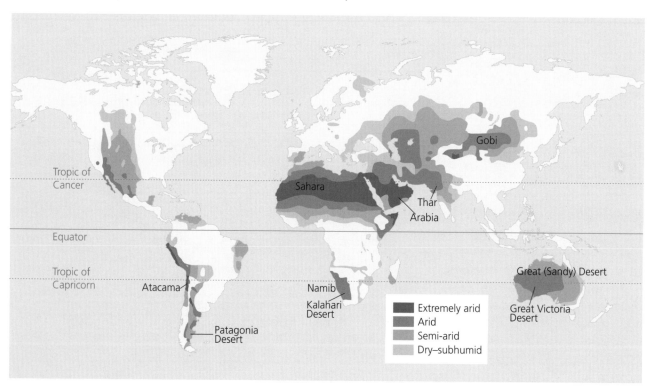

Figure 10.1 The global distribution of arid areas

Many definitions of aridity are based on the concept of **water balance** – i.e. the relationship that exists between inputs in the form of precipitation (*P*) and the losses arising from evaporation and transpiration (*E*).

The actual amount of evapotranspiration that will occur depends on the amount of water available, hence geographers use the concept of potential evapotranspiration (PE), which is a measure of how much evapotranspiration would take place if there was an unlimted supply of water.

The **index of moisture availability** (Im) is calculated as follows:

$$Im = \frac{100S - 60D}{PE}$$

where PE is potential evapotranpiration, S is moisture surplus and D is moisture deficit, aggregated on an annual basis and taking soil moisture storage into account.

- Semi-arid areas have an Im of between −20 and −40.
- Arid areas have an Im of between −40 and −56.
- Extremely arid areas have an Im of less than −56.

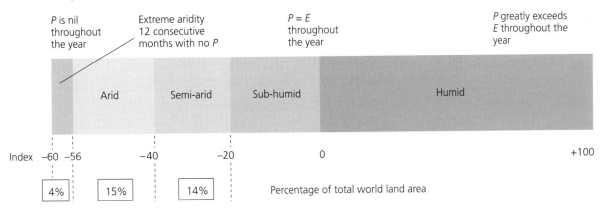

Figure 10.2 The index of aridity

Rainfall effectiveness (P − E) is influenced by a number of factors:

- rate of evaporation – this is affected by temperature and wind speed; in hot, dry areas evaporation losses are high
- seasonality – winter rainfall is more effective than summer rainfall because evaporation losses are lower
- rainfall intensity – heavy, intense rain produces rapid runoff with little infiltration
- soil type – impermeable clay soils have little capacity to absorb water, whereas porous, sandy soils may be susceptible to drought

Causes of aridity

Arid conditions are caused by a number of factors:

- The main cause is the global atmospheric circulation. Dry, descending air associated with the **sub-tropical high pressure** belt is the main cause of aridity around 20°–30°N (Figure 10.3a).
- Distance from sea, **continentality**, limits the amount of water carried across by winds (Figure 10.3b).
- In other areas, such as the Atacama and Namib deserts, **cold off-shore currents** limit the amount of condensation in the overlying air (Figure 10.3c).
- Others are caused by intense **rain shadow effects**, as air passes over mountains (Figure 10.3d).

Typical mistake

Some students confuse aridity and drought. **Aridity** is a permanent water deficit whereas **drought** is an unexpected short-term shortage of available moisture.

Expert tip

There are many variations within deserts – some are sandy, some are stony, some hot, some cold, and they have differing amounts of rainfall. Be careful not to generalise.

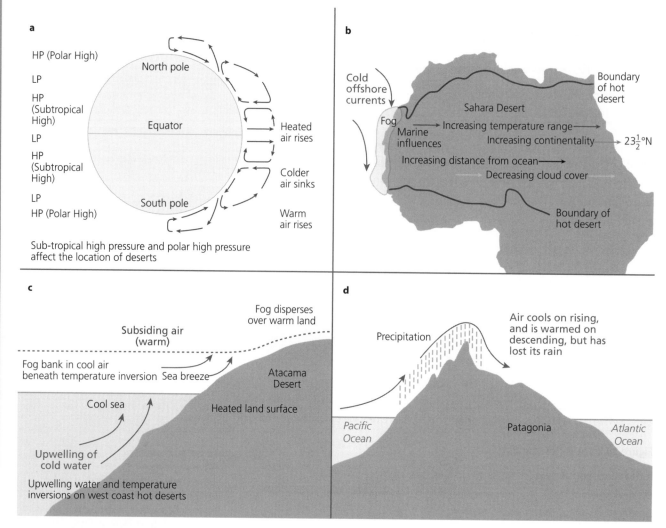

Figure 10.3 The causes of aridity

Now test yourself

Tested

1 Explain the term 'rainfall effectiveness'.
2 Briefly explain why there are deserts on the west coast of southern Africa and the west coast of South America.
3 Explain the main causes of aridity.

Answers on p.219

Desert climates

Revised

Rainfall

The main characteristic of deserts is very low rainfall totals. Desert rain is also highly variable. The inter-annual variability (V) is expressed by:

$$V\ (\%) = \frac{\text{mean deviation from the average}}{\text{average rainfall}} \times 100$$

Variablity in the Sahara is commonly 80–150% compared with just 20% in temperate humid areas. However, many desert areas receive low-intensity rainfall. In coastal areas with cold offshore currents, the formation of fog can provide significant amounts of moisture.

Temperature

Deserts exhibit a wide variation in temperature. Continental interiors show extremes of temperature, both seasonally and diurnally. In contrast, coastal areas have low seasonal and diurnal ranges.

Winds

Deserts are high-wind-energy environments. This is partly due to the lack of vegetation and so less friction between the wind and the surface.

Climate change in deserts

During the Pleistocene Ice Age high latitudes contained more ice (30% of the world surface) than today (10% of the world surface), whereas low-latitude deserts experienced prolonged periods of increased rainfall, known as a pluvials. Some deserts, however, received less rainfall – these dry phases are known as interpluvials. There is widespread evidence for pluvial periods in deserts:

- shorelines marking higher lake levels around dry, salty basins
- fossil soils associated with more humid climates, including horizons containing laterite
- spring deposits of lime, called tufa, indicating higher groundwater levels
- river systems now blocked by sand dunes
- animal and plant remains in areas that are now too arid to support such species
- evidence of human habitation, including cave paintings

The evidence for interpluvials includes sand dunes systems in areas that are now too wet for sand movement to occur. Dunes cannot develop to any great degree in continental interiors unless the vegetation cover is sparse enough to allow sand movement. Satellite imagery and aerial photographs have shown that some areas of forest and savanna, with 750–1500 mm of rain, contain areas of ancient degraded dunes.

Now test yourself

4 Explain the term rainfall variability.

Answer on p.219

Tested

10.2 Processes producing desert landforms

Weathering
Revised

Weathering in deserts is superficial and highly selective. It is greatest in shady sites and in areas within reach of soil moisture; water is important for mechanical weathering, especially exfoliation. Chemical weathering is important and is enhanced in areas that experience dew or coastal fog.

Salt crystallisation causes the decomposition of rock by solutions of salt. When water evaporates, salt crystals are left behind. As the temperature rises, the salts expand and exert pressure on rock. Sodium sulfate (Na_2SO_4) and sodium carbonate (Na_2CO_3) expand by about 300%. This creates pressure on joints, forcing them to crack.

Thermal fracture refers to the break up of rock as a result of repeated changes in temperature over a prolonged period of time.

Disintegration or **insolation weathering** is found in hot desert areas where there is a large diurnal temperature range. In many desert areas daytime temperatures exceed 40°C whereas night-time ones are little above freezing.

The rocks heat up by day and contract by night. As rock is a poor conductor of heat, stresses occur only in the outer layers. This causes peeling or **exfoliation** to occur. In some instances rocks may be split in two. **Block disintegration** is most likely to result from repeated heating and cooling.

Expert tip

Desert processes do not operate in isolation – weathering, erosion, mass movement and deposition are likely to be affecting the same areas

A more localised effect is **granular disintegration**. This occurs due to certain grains being more prone to expansion and contraction than others – this exerts great pressure on the grains surrounding them and forces them to break off.

Wind action   Revised

Movement of sediment is induced by **drag** and **lift** forces, but reduced by **particle size** and **friction**. Drag results from differences in pressure on the windward and leeward sides of grains in an airflow.

There are two types of wind erosion:

- **Deflation** is the progressive removal of small material, leaving behind larger materials. This forms a stony desert or reg. In some cases, deflation may remove sand to form a deflation hollow.
- **Abrasion** is the erosion carried out by wind-borne particles. They act like sandpaper, smoothing surfaces and exploiting weaker rocks. Most abrasion occurs within a metre of the surface, as this is where the largest, heaviest, most erosive particles are carried. Examples of erosional features carved out by abrasion include yardangs, zeugens and ventifacts.

> **Deflation** is lowering of a surface by wind blowing away loose, unconsolidated material.

Transport
Sand-sized particles (0.15–0.25 mm) are moved by three processes:

- **suspension** – particles light enough to be carried by strong winds
- **saltation** – a rolling sand particle gains sufficient velocity for it to leave the sand surface in one or more 'jumps'
- **surface creep** – larger grains are dislodged by saltating grains

> **Now test yourself**
>
> 5 Outline the ways in which wind can erode desert surfaces.
>
> **Answer on p.219**
>
> Tested

The work of water   Revised

There are a number of sources of water in deserts:

- rainfall may be low and irregular, but it does occur, mostly as low-intensity events, although there are occasional flash floods
- deflation may expose the water table to produce an oasis
- rivers flow through deserts; these can be classified as exotic (exogenous), endoreic and ephemeral:
  - **Exotic** or **exogenous** rivers are those that have their source in another wetter environment and then flow through the desert.
  - **Endoreic rivers** are those that drain into an inland lake or sea.
  - **Ephemeral** rivers are those that flow seasonally or after storms. Often they are characterised by high discharges and high sediment levels. Even on slopes as gentle as 2°, overland flow can generate considerable discharges. This is a result of an impermeable surface (in places), limited interception (lack of vegetation) and rain splash erosion displacing fine particles, which in turn seal off the surface and make it impermeable.

Desert landforms

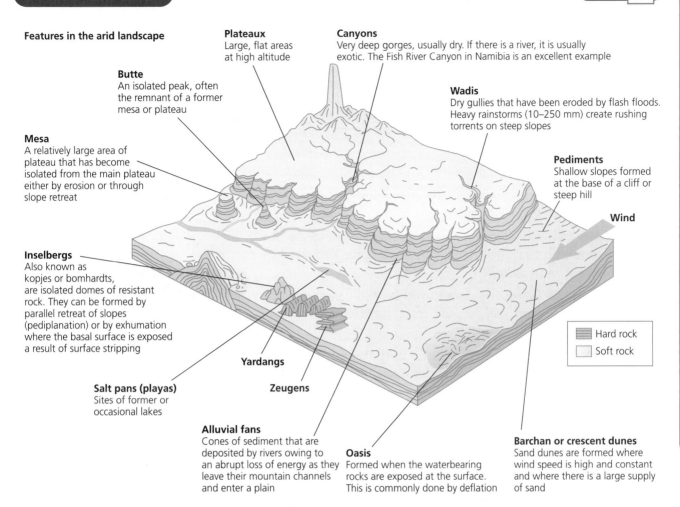

Features in the arid landscape

Plateaux
Large, flat areas at high altitude

Canyons
Very deep gorges, usually dry. If there is a river, it is usually exotic. The Fish River Canyon in Namibia is an excellent example

Butte
An isolated peak, often the remnant of a former mesa or plateau

Wadis
Dry gullies that have been eroded by flash floods. Heavy rainstorms (10–250 mm) create rushing torrents on steep slopes

Mesa
A relatively large area of plateau that has become isolated from the main plateau either by erosion or through slope retreat

Pediments
Shallow slopes formed at the base of a cliff or steep hill

Wind

Inselbergs
Also known as kopjes or bomhardts, are isolated domes of resistant rock. They can be formed by parallel retreat of slopes (pediplanation) or by exhumation where the basal surface is exposed a result of surface stripping

Hard rock
Soft rock

Salt pans (playas)
Sites of former or occasional lakes

Yardangs

Zeugens

Alluvial fans
Cones of sediment that are deposited by rivers owing to an abrupt loss of energy as they leave their mountain channels and enter a plain

Oasis
Formed when the waterbearing rocks are exposed at the surface. This is commonly done by deflation

Barchan or crescent dunes
Sand dunes are formed where wind speed is high and constant and where there is a large supply of sand

Figure 10.4 Desert landforms

Canyons are very deep gorges, and are usually dry. If there is a river it is usually exotic. The Fish River Canyon in Namibia is an excellent example.

Wadis are dry gullies that have been eroded by flash floods. Heavy rainstorms (100–250 mm) create rushing torrents on steep slopes.

Inselbergs are isolated domes of resistant rock. They can be formed by parallel retreat of slopes (pediplanation) or by exhumation, where the basal surface is exposed as a result of surface stripping.

Oases are formed when water-bearing rocks are exposed at the surface. This is commonly done by deflation.

Sand dunes are formed where wind speed is high, constant and there is a large supply of sand.

Pediments are shallow slopes formed at the base of a cliff or steep hill.

Plateaus are large, flat areas at high altitude.

Buttes are isolated peaks, often the remnants of a former mesas or plateaus.

Mesas are relatively large areas of plateau that have become isolated from the main plateau either by erosion or through slope retreat.

Alluvial fans are cones of sediment that are deposited by rivers because of an abrupt loss of energy as they leave their mountain channels and enter a plain.

Yardangs and **zeugens** are wind-eroded landforms where the softer rock strata are removed, leaving the more resistant layers to form either mushroom-shaped features (zeugens) where strata are horizontal, or long ridges (yardangs) where the strata are vertical. These ridges can be as high as 100 m and stretch for many kilometres.

Sand dunes

Winds deposit the sand they carry as dunes. There are many types of dune. Their shape and size depend on the supply of sand, direction of wind, nature of the ground surface, and presence of vegetation.

Only about 25–33% of the world's deserts are covered by dunes and in North America only 1–2% of the deserts are ergs (sandy). Large ergs are found in the Sahara and Arabia.

Types of dune

Nebkhas are small dunes formed behind trees or shrubs, whereas **lunette dunes** are formed in the lee of depressions.

Barchan dunes are crescent-shaped and are found in areas where sand is limited but there is a constant wind supply. They have a gentle windward slope and a steep leeward slope up to 33°.

Parabolic dunes have the opposite shape to barchans – they are crescent shaped but point downwind. They occur in areas of limited vegetation or soil moisture.

Linear dunes or **seifs** occur as ridges 200–500 m apart. They may extend for tens, if not hundreds, of kilometres. It is believed that some regularity of turbulence is responsible for their formation.

Where the winds come from many directions, **star dunes** may be formed, with limbs extending from a central peak. Star dunes can be up to 150 m high and 2 km wide.

> **Typical mistake**
>
> Many students forget that most deserts are stony and that sand dunes only occur in a relatively small number of deserts.

> **Now test yourself**
>
> 6   Distinguish between barchan dunes and parabolic dunes.
>
> 7   Define the terms exogeneous, endoreic and ephemeral with respect to desert rivers.
>
> 8   Explain how (a) alluvuial fans and (b) pediments are formed.
>
> **Answers on p.220**
>
> Tested

10.3 Soils and vegetation

Nutrient flow
Revised

Deserts have low rates of biomass productivity. On average net primary productivity is $90 \, g/m^2/yr$. This is due to the limited amount of organic matter caused by extremes of heat and lack of moisture. Productivity can generally be positively correlated with water availability.

Owing to the low and irregular rainfall, inputs to the nutrient cycle (dissolved in rain and as a result of chemical weathering) are low (Figure 10.5). Most of the nutrients are stored in the soil, and there are very limited stores in the biomass and litter. This is due to the limited amount of biomass and litter in the desert environment. In some deserts nutrient deficiency (especially nitrogen and/or phosphorus) may become critical. The rapid growth of annuals after a rain event rapidly depletes the store of available nutrients, while their return in decomposition is relatively slow.

Despite the extreme short-term variability of the desert environment, the desert ecosystem is considered, in the long-term, to be both stable and resilient. This is due to the adaptations of desert organisms to survive water stress – in some cases for years.

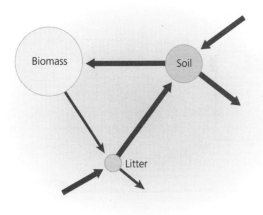

Figure 10.5 Nutrient cycle in a desert

Typical mistake

Deserts are not barren but only 'bloom' periodically following rainfall.

Now test yourself

9 Explain why deserts have low values for NPP (net primary productivity).

Answer on p.220

Tested

Plant and animal adaptations

Revised

Desert plants and animals have acquired similar morphological, physiological and behavioural strategies which, although not unique to desert organisms, are often highly developed and effectively utilised.

The two main strategies are avoidance and tolerance of heat and water stress. The **evaders** comprise the majority of the flora of most deserts. They can survive periods of stress in an inactive state or by living permanently or temporarily in cooler and/or moister environments such as below shrubs or stones, in rock fissures or below ground.

Temperature adaptations

Desert plants and animals are able to function at higher temperatures than their moist-environment counterparts. Some cacti, such as the prickly pear, can survive up to 65°C.

Plants and animals are able to modify the heat of the desert environment in a number of ways:

● Changing the orientation of the whole body can minimise the areas and/or time exposed to maximum heat. Many gazelles, for example, are long and narrow.
● Light colours maximise reflection of solar radiation.
● Surface growths (spines and hairs) can absorb or reflect heat to keep the undersurface cooler, or trap a layer of air, which insulates the underlying surface.
● Body size and shape are especially important in controlling the amount of heat loss – evaporation and metabolism are proportional to the surface area of the plant or animal. The smaller the organism the larger the surface area-to-volume ratio and the greater the heat loss.
● Large desert animals such as the camel can control heating by means of evaporative cooling.

Water loss

To reduce water loss desert plants and animals have many adaptations. Again, a small surface area:volume ratio is an advantage.

Water regulation by plants can be controlled by diurnal closure of stomata, and **xerophytic** plants have a mix of thick, waxy cuticles, sunken stomata and leaf hairs.

Xerophytes are plants that are adapted to drought.

The most drought-resistant plants are the succulents, including cacti, which possess well-developed storage tissues, small surface area-to-volume ratios, and rapid stomatal closure, especially during the daytime.

Now test yourself
Tested

10 Describe the ways in which desert plants are adapted to drought.

Answer on p.220

Desert soils
Revised

Desert soils, called **aridisols**, are only affected by limited amounts of leaching. Soluble salts tend to accumulate in the soil either near the water table or around the depth of moisture percolation. As precipitation declines, this horizon occurs nearer to the surface. Desert soils also have a limited clay content.

Soils with a saline horizon of NaCl (sodium chloride) are called **solonchaks** and those with a horizon of Na_2CO_3 (sodium carbonate) are termed **solonetz**.

> **Aridisols** are desert soils with a low organic content.

Now test yourself
Tested

11 Describe and explain the main characteristics of desert soils.

Answer on p.220

Desertification
Revised

Desertification involves land degradation in non-desert (arid) areas with the loss of biological and economic productivity. It occurs where climatic variability (especially rainfall) coincides with unsustainable human activities. At present, 25% of the global land territory and nearly 16% of the world's population are threatened by desertification.

Causes of desertification

Desertification can be a natural process intensified by human activities. Natural causes of desertification include temporary droughts of high magnitude and long-term climate change towards aridity.

It occurs when already fragile land in semi-arid areas is over-exploited. This overuse can be caused by:

- **over-grazing**, when pastoralists allow too many animals to graze on a fixed area of land
- **over-cultivation**, where the growing of crops exhausts soil nutrients
- **deforestation**, in which too few trees remain after use as firewood to act as windbreaks or to prevent soil erosion

> **Expert tip**
>
> **Desertification** is complex. It can be natural but most contemporary (current) desertification is related to human activities.

> **Desertification** is the spread of desert-like conditions into humid and semi-arid areas.

Now test yourself

12 Suggest a definition for the term 'desertification'.

13 Outline the main natural causes of desertification.

14 Briefly explain two human causes of desertification.

Answers on p.220

Tested

Soil degradation
Revised

Soil degradation includes erosion by wind and water, biological degradation (loss of humus and plant/animal life), physical degradation (loss of structure, changes in permeability) and chemical degradation (acidification, declining fertility, changes in pH, salinisation and chemical toxicity).

> **Soil degradation** is the decline in quantity and quality of soil.

Causes of soil degradation

- Water erosion accounts for about 60% of soil degradation. There are many types of erosion including surface, gully, rill and tunnel.
- Acidification is the change in the chemical composition of the soil, which may trigger the circulation of toxic metals.
- Eutrophication (nutrient enrichment) can degrade the quality of ground water.
- Groundwater over-abstraction can lead to dry soils.
- Salt-affected soils are typically found in marine-derived sediments, coastal locations and hot arid areas where capillary action brings salts to the upper part of the soil.
- Atmospheric deposition of heavy metals and persistent organic pollutants can make soils less suitable for sustaining the original land cover and land use.
- Climate change will probably intensify the problem. It is likely to affect hydrology and hence land use.

Human activities

Human activities have often led to degradation of the world's land resources. Damage has occurred on 15% of the world's total land area. While severely degraded soil is found in most regions of the world, the negative economic impact of soil degradation is most severe in those countries most dependent on agriculture for their incomes.

Table 10.1 Human activities and their impact on soil erosion

Action	Effect
Removal of woodland or ploughing of established pasture	The vegetation cover is removed, roots binding the soil die and the soil is exposed to wind and water. Slopes are particularly susceptible to erosion.
Cultivation	Exposure of bare soil surface before planting and after harvesting. Cultivation on slopes can generate large amounts of runoff and create rills and gullies.
Grazing	Overgrazing can severely reduce the vegetation cover and leave the surface vulnerable to erosion. Grouping of animals can lead to trampling and creation of bare patches. Dry regions are particularly susceptible to wind erosion.
Roads or tracks	Reduced infiltration can cause rills and gullies to form.
Mining	Exposure of bare soil leads to rapid erosion.

10.4 Sustainable management of arid and semi-arid environments

Sustainable development is a process by which human potential (quality of life) is improved and the environment (resource base) is used and managed sustainably to supply humanity on a long-term basis.

> **Sustainable development** meets the needs of the present generation without compromising the ability of future generations to meet their own needs.

Essential oils in the Eastern Cape, South Africa
Revised

Global pattern

About 65% of the world production of essential oils occurs in developing countries such as India, China, Brazil, Indonesia, Mexico, Egypt and Morocco. However, the USA is also a major producer of essential oils such as peppermint and other mints.

Globally, the essential oils industry – valued at around $10 billion – is enjoying huge expansion. Opportunities include increasing production of existing products, and extending the range of crops grown.

Essential oils in South Africa

The South African essential oils industry has only recently emerged. Currently it exports mainly to developed economies such as Europe (49%), USA (24%) and Japan (4.5%). The most significant essential oils produced by South Africa are eucalyptus, citrus, geranium and buchu. Others include rosemary and lavender.

Currently, the South African essential oils industry comprises about 100 small commercial producers of which fewer than 20% are regular producers. Expansion of the industry in the area involves much-needed agricultural and agri-processing diversification.

Factors that make South Africa an attractive essential oils market include:
- much of the demand being in the northern hemisphere, with seasonal effects making southern hemisphere suppliers globally attractive
- having traditionally strong trade links with Europe as a major importer of fragrance materials
- South Africa being established as a world-class agricultural producer of a wide range of products

The Eastern Cape

The Eastern Cape is set to become one of the main contributors to South Africa's burgeoning essential oils industry, with 10 government-sponsored trial sites currently in development throughout the province. Six of these form part of the Essential Oil Project of Hogsback, where approximately eight hectares of communal land is being used. A project at Keiskammahoek has been operational since 2006. These trials form part of a strategy to develop a number of essential oil clusters in the Eastern Cape.

The production of essential oils holds considerable potential as a form of sustainable agricultural development in the Eastern Cape. Not only are the raw materials present but it is a labour-intensive industry and would utilise a large supply of unemployed and underemployed people.

The essential oils industry has a number of advantages:
- It is a new or additional source of income for many people.
- It is labour intensive and local in nature.
- Many plants are already known and used by the peoples as medicines, and are therefore culturally acceptable.
- In their natural state the plants are not very palatable nor of great value and will not therefore be stolen.
- Many species are looked upon as weeds – removing these regularly improves grazing potential as well as supplying raw materials for the essential oils industry.

Crops

Wild als (*Artemisia afra*) is an indigenous mountainous shrub, used for the treatment of colds. Its oil has a strong medicinal fragrance and is used in deodorants and soaps. Double cropping – in summer when the plant is still growing and in autumn at the end of the growing season – yields the best results.

Demand for *Artemisia* has not outstripped the supply of wild material. However, it is increasingly being cultivated as a second crop. It requires minimal input in terms of planting, tillage, pest control etc. and it is relatively easy to establish and manage. Moreover, it can stabilise many of the maize fields and slopes where soil erosion is a problem.

> **Expert tip**
>
> You are expected to know a case study of sustainable management in arid and semi-arid areas. You will need to learn some supporting details about your case study.

The local population is very enthusiastic about the crop, especially given the right economic incentives.

Khakibush (*Tagetes minuta*) is an aromatic. In the Eastern Cape it is the most common weed in most maize fields. Oil of tagetes is an established essential oil, although its market is limited.

Local people are again quite enthusiastic about collecting khakibush if the incentives are there. Harvesting takes up to 3 months and provides a great deal of extra employment, as well as eradicating a weed.

At present the use of wild *Tagetes* and those in the maize fields is sufficient to meet demand. An increase in demand might lead to the establishment of *Tagetes* as secondary crop in maize fields.

Other species such as geranium, peppermint and sage require too much land, labour and water to be very successful.

Typical mistake

Sustainable development is an ongoing process. It is possible to evaluate its success so far – or its potential for success – but we cannot know how conditions will change for future generations.

Now test yourself

15 To what extent could the essential oils industry be considered a form of sustainable development in the Eastern Cape.

Answer on p.220

Tested

Exam-style questions

1 **(a)** Describe the distribution of the world's arid areas. [10]

  (b) Evaluate the role of wind in hot desert environments. [15]

2 **(a)** Explain the causes of aridity. [10]

  (b) Outline the potential for sustainable development in dry areas. [15]

Exam ready

11 Production, location and change

11.1 Agricultural systems and food production

Factors affecting agricultural land use

Revised

Physical factors

Physical factors set broad limits as to what can be produced. The farmer's decisions are then influenced by economic, social/cultural and political factors. Key physical factors are:

- temperature
- precipitation
- soil type and fertility

Locally, aspect, angle of slope and wind intensity may also be important factors in deciding how to use the land.

Cotton, for example, needs a frost-free period of at least 200 days. Rainfall should be over 625 mm a year with not more than 250 mm in the autumn harvest season. Cotton production is now highly mechanised. **Irrigation** has allowed cotton to flourish in the USA in the drier western states. In contrast, cotton acreage has fallen considerably in the southern states. A crop pest called the cotton boll weevil, which caused great destruction to cotton crops in the past, has been a big factor in the diversification of agriculture in the southern states.

Water is vital for agriculture. Irrigation is an important factor in farming in many parts of the world. There is a divide by world region between those relying on **rainfed water** for crop use and those needing irrigation. The highest proportion of irrigation water use is in the Middle East, North Africa and South Asia. Table 11.1 compares the main types of irrigation – an example of the ladder of agricultural technology.

> **Irrigation** supplies dry land with water using systems of ditches and also by more advanced means.
>
> **Rainfed water** is water for crops that comes from precipitation as opposed to irrigation.

Table 11.1 Main types of irrigation

Type	Efficiency (%)
Surface – used in over 80% of irrigation worldwide	
Furrow – traditional method; cheap to install; labour-intensive; high water losses; susceptible to erosion and salinisation	20–60
Basin – cheap to install and run; needs a lot of water; susceptible to salinisation and waterlogging	50–75
Aerial (sprinklers) – used in 10–15% of irrigation worldwide	
Costly to install and run; low-pressure sprinklers preferable	60–80
Sub-surface ('drip') – used in 1% of irrigation worldwide	
High capital costs; sophisticated monitoring; very efficient	75–95

Economic factors and agricultural technology

Economic factors include transport, markets, capital and technology. The cost of growing different crops or keeping different livestock varies. The market prices for agricultural products will vary also and can change from year to year. The necessary investment in buildings and machinery can mean that some changes in farming activities are very expensive. In most countries there has been a trend towards fewer but larger farms. Large farms allow **economies of scale** to operate.

Distance from markets has always been an important influence on farming. Von Thunen argued that the price a farmer obtained for a unit of his product was equal to its price at market less the cost of transporting it to the market. He called this **economic rent** (Figure 11.1). Thus the nearer a farmer was to the market the greater his returns from the sale of his produce. Also, land closest to the market would be the most intensively farmed.

> **Economies of scale** refer to the reduction in unit cost as the scale of an operation increases.
>
> **Economic rent** is the profit from a unit of land after transport costs to market are deducted.

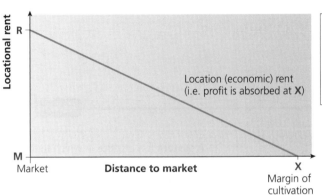

R = locational rent
M = Market
X = Margin of cultivation (here the cost of transport to the market is so great that it equals the profit that can be made by selling produce in the market)

Figure 11.1 The relationship between economic rent and distance from the market

The development and application of **agricultural technology** requires investment. The status of a country's agricultural technology is vital for its food security and other aspects of its quality of life. An important form of aid is the transfer of agricultural technology from more advanced to less advanced nations.

Agro-industrialisation or industrial agriculture refers to the industrialised production of livestock, poultry, fish and crops. Its characteristics include:

- very large farms
- concentration on one (monoculture) or a small number of farm products
- a high level of mechanisation
- low labour input per unit of production
- heavy usage of fertilisers, pesticides and herbicides
- sophisticated ICT management systems

> **Agricultural technology** is technology used to control the growth and harvesting of animal and vegetable products.
>
> **Capital-intensive farming** involves systems where the cost of inputs other than labour is high relative to the labour input.

This type of large-scale, **capital-intensive farming** originally developed in Europe and North America and then spread to other parts of the developed world. It has been spreading rapidly in many developing countries since the beginning of the Green Revolution. Examples of agro-industrialisation include the Canadian Prairies and the Mato Grosso in Brazil.

Political factors

The influence of government on farming has steadily increased in many countries. For example, in the USA the main parts of federal farm policy over the past half century have been:

- price support loans
- production controls
- income supplements

> **Typical mistake**
>
> Students often write as if agro-industrialisation is confined to developed countries, when large areas of some developing countries, such as the Pampas in Argentina, are characterised by this type of agriculture. This is a consequence of the globalisation of agriculture.

Thus, the decisions made by individual farmers are heavily influenced by government policies. However, in centrally planned economies the state has far more control. This was the case for many years in the former Soviet Union and China. An agricultural policy can cover more than one country, as evidenced by the EU's Common Agricultural Policy.

Social/cultural factors

There is a tendency for farmers to stay with what they know best and often a sense of transgenerational responsibility to maintain family farming tradition. Legal rights and **land tenure** are heavily influenced by culture:

- In the past, inheritance laws have had a huge impact on the average size of farms.
- In some countries it has been the custom on the death of a farmer to divide the land equally between all his sons, but rarely between daughters.
- The reduction in the size of farms by these processes often reduced them to operating at only a subsistence level.
- In most societies women have very unequal access to, and control over, rural land and associated resources.

> **Land tenure** regulates the ways in which land is, or can be, owned.

Agricultural systems

Revised

Farms can be seen to operate as a **system**. A farm requires a range of inputs such as labour and energy so that the processes (throughputs) that take place on the farm can be carried out. The aim is to produce the best possible outputs, such as milk, eggs, meat and crops. A profit will only be made if the income from selling the outputs is greater than expenditure on the inputs and processes. Different types of agricultural system can be found within individual countries and around the world. The most basic distinctions are between:

- arable, pastoral and mixed farming
- subsistence and commercial farming
- **extensive** and **intensive farming**
- organic and non-organic farming

> A **system** is a practice in which there are recognisable inputs, processes and outputs.
>
> **Extensive farming** is where a relatively small amount of agricultural produce is obtained per hectare of land, so such farms tend to cover large areas. Inputs per unit of land are low.
>
> **Intensive farming** is characterised by high inputs per unit of land to achieve high yields per hectare.

Case study A pastoral system: sheep farming in Australia

Sheep farming in Australia is a classic example of extensive farming.

- The main physical input is the extensive use of natural open ranges, often fragile in nature. Human inputs are low compared with most other types of agriculture with very low use of labour and capital per hectare.
- The main processes are grazing, lambing, dipping and shearing.
- The outputs are lambs, sheep, wool and sheep skins.

Australia is the world's leading sheep producing country with a total of about 120 million sheep. 75% of the country's sheep are Merinos, which produce very high-quality clothing wool. Sheep and wool production occurs in three geographical zones:

- the high-rainfall coastal zone
- the wheat/sheep intermediate zone
- the pastoral interior zone

In the coastal and intermediate zones the best land is reserved for arable farming, dairy and beef cattle and market gardening. Sheep are frequently kept on the more marginal areas. This type of agriculture is at its most extensive in the pastoral zone, which is the arid and semi-arid inland area.

The main issues in Australian sheep farming are:
- weed infestation
- destruction of wildlife habitats due to sheep grazing
- the occurrence of periodic droughts
- soil loss from wind erosion and loss of soil structure
- animal welfare
- increasing concern about the shortage of experienced sheep shearers

Location

- Temperatures of 21°C and over throughout the year, allowing two crops to be grown annually. Rice needs a growing season of only 100 days.
- Monsoon rainfall over 2000 mm, providing sufficient water for the fields to flood, which is necessary for wet rice cultivation.
- Rich alluvial soils built up through regular flooding.
- An important dry period for harvesting the rice.

A water-intensive staple crop

Production systems are extremely water intensive. Much of Asia's rice production is intensive subsistence cultivation where the crop is grown on very small plots of land using a very high input of labour. 'Wet' rice is grown in the fertile silt and flooded areas of the lowlands while 'dry' rice is cultivated on terraces on the hillsides.

The farming system

Padi-fields characterise lowland production. At first, rice is grown in nurseries, then transplanted when the **monsoon rains** flood the padi-fields. The main rice crop is harvested when the drier season begins in late October. A second rice crop can then be planted in November.

Water buffalo are used for work. The labour-intensive nature of rice cultivation provides work for large numbers of people. A high labour input is needed to:

- build the embankments that surround the fields
- construct irrigation canals
- plant nursery rice, plough the padi-field, transplant the rice from the nursery to the padi-field, weed, and harvest the mature rice crop
- cultivate other crops in the dry season and possibly tend a few chickens or other livestock

Rice seeds are stored from one year to provide the next year's crop.

Issues in the intensification of agriculture and the extension of cultivation

Revised

Higher agricultural production can be achieved in two ways:

- increasing the land under cultivation
- increasing the yield per hectare

The intensification of agriculture has occurred through the use of high-yielding crop varieties, fertilisers, herbicides and pesticides, and irrigation. The industrialised farmlands of today are all too frequently lacking in the wildflowers, birds and insects of the past. These sterilised landscapes provide relatively cheap food, but at high environmental cost.

The intensification of agriculture can result in **soil degradation**. This is a global process. It impacts significantly on agriculture and also has implications for the urban environment, pollution and flooding. Research has shown that the heavy and sustained use of artificial fertiliser can result in serious soil degradation.

The environmental impact of the Green Revolution

Much of the global increase in food production in the last 50 years can be attributed to the **Green Revolution**. Although the benefits of the Green Revolution are clear, serious criticisms have also been made:

- The necessary high inputs of fertiliser and pesticide are costly in both economic and environmental terms.
- The problems of salinisation and waterlogged soils have increased with the expansion of irrigation.
- High chemical inputs have had a considerable negative effect on biodiversity.
- Ill-health has occurred due to contaminated water and other forms of agricultural pollution.
- Green Revolution crops are often low in important minerals and vitamins.

> **Soil degradation** is the physical loss (erosion) and the reduction in quality of topsoil associated with nutrient decline and contamination.

> **Expert tip**
>
> The intensification of agriculture and the extension of cultivation have been very important in increasing food production, but along with the obvious advantages of such processes have come significant environmental consequences. It is important to be aware of both the advantages and disadvantages of any major processes such as these.

> The **Green Revolution** has involved the introduction of high-yielding crops and modern agricultural techniques in developing countries.

Now test yourself

1 What are the characteristics of agro-industrialisation?
2 Distinguish between intensive and extensive farming.
3 Distinguish between wet rice and dry rice.
4 What are the two ways in which agricultural production can be increased?
5 What have been the main problems of the Green Revolution?

Answers on p.220

11.2 The management of agricultural change: Jamaica

The importance of agriculture

Agriculture in Jamaica is dominated by the production of **traditional crops** such as sugar, bananas and coffee. In addition, a number of **non-traditional crops** including sweet potatoes, yams and hot peppers are cultivated. Over the past two decades the major export earner, sugar, has experienced a considerable decline. Both sugar and bananas in particular have had to contend with price and market insecurity as a result of preference erosion in the EU market. Agriculture contributes 7% to Jamaica's GDP and employs about 20% of the workforce.

Recent changes in Jamaican farming
Significant areas of **marginal land** were abandoned in recent years because of:

● the difficulties of making a living on marginal land
● the removal of **preferential treatment** for bananas on the European market
● crop disease

Climatic hazards often impact substantially on farming in Jamaica – hurricanes in particular, but also drought. Jamaica is also having to address the issue of land degradation.

Policy responses
In response the Jamaican government announced a new policy for a sustainable local sugar industry. Commodity-specific policies for bananas and cocoa were also introduced. Jamaica has also produced a New Agricultural Development Plan that aims to transform the farming sector by 2020. The main objectives of the plan are to:

● halt the decline of the agricultural sector
● restore productivity to agricultural resources
● ensure that farming communities provide meaningful livelihoods and living environments for those who depend on the agricultural sector

As exports of some traditional farm products have declined the Jamaican government has tried to encourage **agricultural diversification**. The exploitation of **niche markets** has been a major aspect of the modernisation of Jamaican agriculture. The government has recognised the contribution ICT can make to enhancing the sector's efficiency and productivity. The range of policies introduced by the government in recent years has undoubtedly helped to bring about beneficial changes in Jamaican agriculture.

Traditional crops are major crops that have been cultivated over a long period of time.

Non-traditional crops are newer crops cultivated on a reasonably large scale, which mark the diversification of agricultural production.

Marginal land is land of poor quality because of lack of nutrients, soil erosion, distance from markets and other human and physical factors.

Preferential treatment refers to former colonies of European Union countries being given preferential access to EU markets for a certain time period so their economies could adjust to EU trade policies.

Typical mistake

Students can sometimes struggle with the concept of marginal land. Land can be classed as marginal for a number of reasons, but the essence of the concept is that the returns from farming such land are very limited (marginal) compared with better-quality farmland elsewhere.

Agricultural diversification is the process by which countries and individual farms expand the range of their activities.

A **niche market** is a subset of a market on which a specific product is focused.

Kew Park Farm is a mixed commercial farm in the west of Jamaica (Figure 11.2). This is a very hilly part of Jamaica and good management has been essential for the farm's survival.

Most of the farmed area is allocated to beef cattle with a total of about 700 animals. Other parts of the farm support a variety of agricultural activities:

● An area of 16 hectares is planted with arabica coffee.
● Two hectares are given over to citrus fruits (ortaniques).
● There are two hectares of lychees.
● The farm supports some 2000 free-range chickens.
● Five pig units comprise a total of 120 sow units (breeding animals) and 2500 fatteners.

Life in rural Jamaica is not easy and Kew Park provides the only full-time employment in the area. Wages are low. Farm managers have had to be constantly aware of the costs of all their inputs and processes. Knowledge of local and more distant markets in terms of both access and price are important. The farm also has to be aware of government agricultural policies and incentives. The damage caused by a pest known as the coffee berry borer can eliminate the profit expected from a coffee crop in Jamaica.

The farm's website highlights 'Kew Park Essentials', a range of traditional herbal remedies, spicy foods and refreshers. Although production of these products is not new, the marketing of them has changed considerably in recent years. This aspect of the farm's production has accounted for an increasing proportion of its income in recent years.

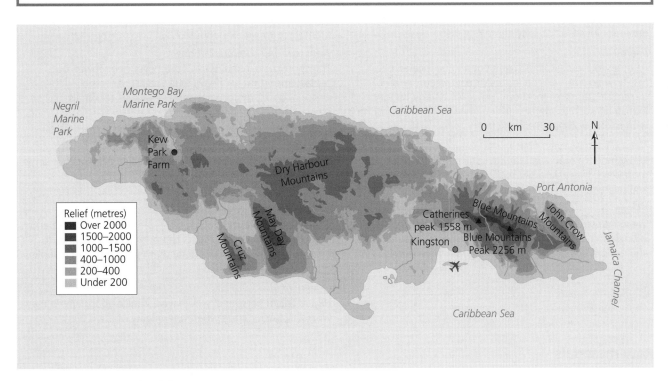

Figure 11.2 Jamaica: relief and drainage

Now test yourself

Tested

6 What is the climate of Jamaica?
7 Why have significant areas of marginal land been abandoned?
8 How has farming in Jamaica become more diversified?
9 Name the general type of farming practised on Kew Park Farm.
10 What are 'Kew Park Essentials'?

Answers on pp.220–221

Expert tip

A simple, but clearly labelled sketch map can considerably enhance the presentation of a case study. With practice you should be able to draw such a sketch map in about a minute. Use Figure 11.2 to draw a simple sketch map showing the location of Kew Park farm.

11.3 Manufacturing and related service industry

Industrial location: influential factors

Table 11.2 outlines the important factors controlling industrial location.

Table 11.2 Factors controlling the location of industry

Physical factors	Human factors
Site: the availability and cost of land Large factories need flat, well-drained land on solid bedrock. An adjacent water supply may be essential.	*Capital (money)* Business people, banks and governments are more likely to invest money in some areas than others.
Raw materials Industries requiring heavy and bulky raw materials will generally locate as close to these raw materials as possible.	*Labour* The quality and cost of labour are key factors. Reputation, quantity, turnover and mobility can also be important.
Energy At times in the past, industry needed to be located near fast-flowing rivers or coal mines. Today, electricity can be transmitted to most locations. However, energy-hungry industries, such as metal smelting, may be drawn to countries with relatively cheap hydroelectricity.	*Transport and communications* Transport costs in real terms have never been lower, but remain important for heavy, bulky items. Accessibility to airports, ports, motorways and key railway terminals may be crucial factors for some industries.
Natural routeways and harbours Many modern roads and railways still follow natural routeways. Natural harbours provide good locations for ports and related industrial complexes.	*Markets* The location and size of markets is a major influence for some industries.
Climate Some industries such as aerospace benefit directly from a sunny climate. Indirect benefits such as lower heating bills and a more favourable quality of life may also occur.	*Government influence* Government policies and decisions can have a big direct and indirect impact. Governments can encourage industries to locate in certain areas and deny them planning permission in others.
	Quality of life Highly skilled personnel who have a choice about where they work will favour areas where the quality of life is high.

Raw materials

Industries that use raw materials directly are known as **processing industries**. The processes involved in turning a raw material into a manufactured product usually result in **weight loss**. If weight loss is substantial the location of the factory will be drawn towards the raw material that is most costly to transport. **Tidewater locations** are particularly popular with industries using significant quantities of imported raw materials, such as flour milling and oil refining. Tidewater locations are **break-of-bulk** points.

Markets

Where the cost of distributing the finished product is the greater part of total transport costs a market location is logical. A small number of industries, including soft drinks and brewing, are 'weight gaining' and are thus market oriented. However, there are other reasons for market location. Industries where fashion and taste are variable need to be able to react quickly to changes demanded by their customers.

Energy

The **Industrial Revolution** was based on the use of coal, which is costly to transport. Thus, industrial towns and cities developed on coalfields, which became focal points for the developing transport networks. The investment in

Weight loss refers to industries for which the finished product is lighter than the weight of the raw materials required to manufacture it.

Tidewater locations are port locations where freight can be transferred between road, rail and pipeline to water transport.

Break of bulk refers to a location, such as a seaport, where freight has to be transferred from one mode of transport to another.

The **Industrial Revolution** involved the transformation in the late eighteenth and nineteenth centuries of first Britain and then other European countries and the USA from agricultural into industrial nations.

both **hard** and **soft infrastructure** was massive, so that even when new forms of energy were substituted for coal, many industries remained at their coalfield locations, a phenomenon known as **industrial inertia**. During the twentieth century the construction of national electricity grids and gas pipeline systems made energy virtually a ubiquitous resource in the developed world. As a result most modern industry is described as **footloose**.

Transport

The share of industry's total costs accounted for by transportation has fallen steadily over time due to:

- major advances in all modes of transport
- great improvements in the efficiency of transport networks
- technological developments moving industry to the increasing production of higher-value/lower-bulk goods

The cost of transport has two components: **fixed (terminal) costs** and **line-haul costs**. While water and pipeline transport have higher fixed costs than rail and road, their line-haul costs are significantly lower. Air transport, which suffers from both high fixed and line-haul costs, is only used for high-value freight or for goods, such as flowers, that are extremely perishable.

Land

Technological advance has made modern industry much more space-efficient. However, modern industry is horizontally structured (on one floor) as opposed to, for example, the textile mills of the nineteenth century with four or five floors. In the modern factory transportation (car parks etc.) takes up much more space than it used to.

During the Industrial Revolution entrepreneurs had a relatively free choice of where to locate in terms of planning restrictions. However, over time more and more areas have been placed off limits to industry.

Capital

That part of **capital** invested in plant and machinery is known as fixed capital, as it is not mobile compared with working capital (money). In the early days of the Industrial Revolution the availability of capital was geographically constrained. This was a significant factor in the clustering of industry. This factor has a minimal constraining influence in the developed world today. However, in LEDCs the constraints of capital are usually much greater. Virtually all industries have over time substituted capital for labour in an attempt to reduce costs.

Labour

Although all industries have become more capital intensive over time, labour still accounts for over 20% of total costs in manufacturing industry. The cost of labour can be measured in two ways: as wage rates and as unit costs. The former are the hourly or weekly amounts paid while the latter are a measure of productivity, relating wage rates to output. Certain skills sometimes become concentrated in particular areas – the **sectoral spatial division of labour**.

- Variation in wage rates can be identified at different scales. By far the greatest disparity is at the global scale.
- There are wide variations in non-wage labour costs.
- The availability of labour as measured by high rates of unemployment is not an important location factor.
- Both the **geographical mobility of labour** and the occupational mobility of labour are limited.
- The reputation of a region's labour force can influence inward investment.

Hard infrastructure involves the basic utilities (e.g. road, rail, air links, water, sewerage and telephone systems) which provide a network that benefits business and the community.

Soft infrastructure involves other services such as health, education, banking and retailing that are important to business and the community.

A **footloose industry** is one that is not tied to certain areas because of its energy requirement or other factors.

Fixed (terminal) costs are those accrued by the equipment used to handle and store goods, and those of providing the transport system.

Line haul costs are those of actually moving the goods, which are largely composed of fuel costs and wages.

Capital represents the finance invested to start up a business and to keep it in production.

> **Typical mistake**
>
> Students sometimes confuse fixed capital, which is that part of capital invested in plant and machinery, with working capital, which is the money used to pay for regular expenditure on factors such as labour and raw materials.

Sectoral spatial division of labour is the concentration of specific industrial skills in particular regions or countries.

Geographical mobility of labour is the relative ability of labour to move from one geographical region to another.

Economies of scale: internal and external

Economists recognise five types of **internal economies of scale**:

- bulk-buying economies
- technical economies
- financial economies
- marketing economies
- managerial economies

However, it is possible that an increase in production at some stage might lead to rising unit costs. If this happens, **diseconomies of scale** are said to exist (Figure 11.3).

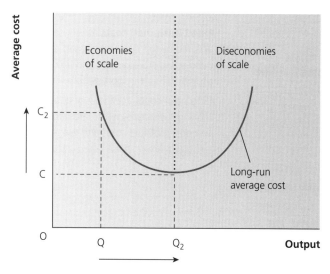

Figure 11.3 Economies and diseconomies of scale

External economies of scale (agglomeration economies) can be subdivided into:

- urbanisation economies
- localisation economies

However, when an urban-industrial area reaches a certain size, urbanisation diseconomies may come into play. For example, high levels of traffic congestion may push up transport costs and intense competition for land will increase land prices and rents.

Government policies

In the old-style communist centrally planned economies the influence of government on industry was absolute. In other countries the significance of government intervention has depended on:

- the degree of public ownership
- the strength of regional policy in terms of restrictions and incentives

Governments influence industrial location for economic, social and political reasons.

Technology

Technology can influence industrial location in two main ways:

- The level of technological development in terms of infrastructure and human skills.
- Technological advance may attract a company or industry to new locations.

Internal economies of scale occur when an increase in production results in a lowering of unit costs.

External economies of scale (agglomeration economies) are the benefits that accrue to a firm by locating in an established industrial area. External economies of scale can be subdivided into urbanisation economies and localisation economies.

Expert tip

Be clear about the distinction between internal and external economies of scale. The former relates to reducing costs because of changes inside a company, while the latter relates to a fall in costs due changes outside of the company.

Case study **Slovakia: the changing location of EU car manufacturing**

Investment in car manufacturing has shifted from western to eastern Europe in recent years as countries like Slovakia have joined the EU. Slovakia can manufacture cars at a lower cost due to:

- relatively low labour costs and company taxation rates
- a highly skilled workforce, particularly in areas once important for heavy industry
- a strong work ethic resulting in high levels of productivity
- low transport costs because of proximity to western European markets

- very low political risk because of the stable nature of the country
- attractive government incentives
- good infrastructure in and around Bratislava and other selected locations
- an expanding regional market for cars as per capita incomes increase

Major car manufacturers operating in Slovakia include Volkswagen, Hyundai and Peugeot.

Industrial agglomeration and functional (industrial) linkages

Revised

Industrial agglomeration can result in companies enjoying the benefits of external economies of scale and industrial (functional) linkages. Industrial linkages are the contacts and flows of information between companies that can happen more cheaply and easily when companies locate in close proximity. Three types of linkage are generally recognised: backward, forward and horizontal.

Industrial estates

Industrial estates can be found in a range of locations from inner cities to rural areas. The logic behind industrial estates includes:

- concentrating in a delimited area to reduce the per-business expense
- attracting new business by providing an integrated infrastructure in one location
- separating industrial uses from residential areas
- providing for localised environmental controls
- eligibility of industrial estates for grants and loans under regional development policies

Export processing zones

There are a number of different types of **export processing zones** (EPZs) which include free trade zones and free ports. EPZs have evolved from initial assembly and simple processing activities to include high-tech and science parks, finance zones, and logistics centres.

> **Industrial agglomeration** is the clustering together and association of economic activities in close proximity to one another.
>
> An **industrial estate** is an area zoned and planned for the purpose of industrial development.
>
> **Export processing zones** are industrial zones with special incentives set up to attract foreign investors, in which imported materials undergo some degree of processing before being re-exported.

Formal and informal sectors of employment

Revised

Formal sector jobs generally provide better pay and much greater security. Fringe benefits such as holiday and sick pay may also be available. Formal sector employment includes government workers, and people working in established manufacturing and retail companies.

Employment in the **informal sector** is generally low paid and often temporary and/or part-time. While such employment is outside the tax system, job security is poor, with an absence of fringe benefits. About three-quarters of those working in the informal sector are employed in services. Typical jobs are shoe-shiners and market traders. The size of the informal labour market varies from an estimated 4–6% in the high-income countries to over 50% in the low-income countries.

> **Formal sector** jobs are known to the government department responsible for taxation and to other government offices.
>
> The **informal sector** refers to that part of the economy operating outside official government recognition.

The advantages of the informal sector (when there are not enough formal sector jobs) are that it:

- provides jobs
- alleviates poverty
- bolsters entrepreneurial activity
- facilitates community cohesion and solidarity

Now test yourself Tested

11 Distinguish between fixed transport costs and line-haul costs.
12 Distinguish between wage rates and unit costs.
13 What are external economies of scale?
14 What is industrial agglomeration?
15 What are export processing zones?

Answers on p.221

11.4 The management of industrial change: India

India is a newly industrialised country where the transformation of the economy has been based more on the service sector than on manufacturing. This has been at least partly due to a low level of **foreign direct investment** in manufacturing, a situation that began to change in the early 1990s. The service sector accounts for 52% of India's GDP, with manufacturing industry and agriculture respectively responsible for 28% and 17% (Figure 11.4).

- Textiles is the largest industry in the country, employing about 20 million people and accounting for one-third of India's exports.
- The car industry has expanded significantly in recent times and is now the seventh largest in the world.
- However, it is in the field of software and ICT in general that India has built a global reputation.

> **Foreign direct investment** is overseas investment in physical capital by transnational corporations.

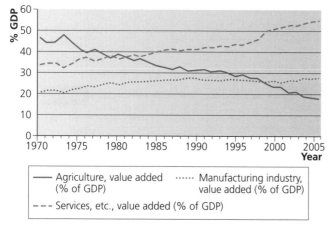

Figure 11.4 The main economic sectors' share of Indian GDP

Key:
— Agriculture, value added (% of GDP)
······ Manufacturing industry, value added (% of GDP)
--- Services, etc., value added (% of GDP)

> **Typical mistake**
>
> Students should take care with the use of the word 'industry' as it can be applied to all sectors of the economy, for example the agricultural industry and the service industry. If you use it with reference to the manufacture of goods then state clearly that this is 'manufacturing industry'.

Traditional industrial policy

The objective of India's policy from the 1950s was to coordinate investment decisions in both the public and private sectors and to bring certain strategic industries under public ownership. Five-year plans were set up and this state-directed industrialisation model was followed from 1950 to 1980. The main objectives of the plans were:

- industrialisation
- raising per capita incomes
- achieving equity in the distribution of gains from economic development
- reducing the existing concentration of economic power
- achieving a more even regional distribution of industrial development

The role of **heavy industry**, particularly iron and steel, was emphasised, with the public sector playing a major role. Technological self-reliance became an important element of industrial policy. Industrial policy measures under the 5-year plans included: industrial licensing, strict import controls, subsidising exports, and strict controls on investment by transnational corporations. The range of controls made India one of the most protected economies in the world. High tariffs made imports very expensive and thus controlled their volume.

> **Heavy industry** is one with heavy/bulky raw materials and heavy/bulky finished products.

Economic reform

The currency crisis of 1991 proved to be a major turning point, instigating bold economic reforms that resulted in rapid economic growth. The reforms were based on:

- liberalisation
- deregulation
- market orientation

Tariffs on imports were significantly reduced along with other non-tariff trade barriers as a result of India's membership of the World Trade Organization (WTO).

In the last two decades India has spawned a modern, highly export-oriented ICT industry. The export intensity of Indian software is more than 70% and the ICT sector has attracted high levels of foreign direct investment.

India's technological success has not been confined to the ICT industry. India's corporations have achieved significant growth in a number of industries including, in particular, pharmaceuticals and car components. It is one of only three countries in the world to build super-computers on its own and one of only six countries in the world to launch satellites.

Bangalore

Bangalore is the most important individual centre in India for high technology. In the 1980s Bangalore became the location for the first large-scale foreign investment in high technology in India. India's ICT sector has benefited from the filtering down of business from the developed world. Many European and North American companies outsource to Indian companies because:

- labour costs are considerably lower
- a number of developed countries have significant ICT skills shortages
- India has a large and able English-speaking workforce

Regional policy

Like many other countries India adopted regional economic planning and tried to encourage a better spread of industry around the country. In the early 1970s backward states and backward districts were identified and a scheme of incentives for industry to locate in these regions was introduced. This included:

- a grant of 15% of fixed capital investment
- transport subsidies
- income tax concessions

The state of India's infrastructure is an important issue. Infrastructure in India is at a lower level than that in China and other newly industrialised countries in the region.

Now test yourself

Tested

16 What proportions of india's GDP are accounted for by agriculture, industry and services?

17 What three principles were the economic reforms of the early 1990s based upon?

18 Why have many foreign firms outsourced to Indian companies?

Answers on p.221

Exam-style questions

1 (a) Discuss the physical factors that can influence agricultural land use. [10]

(b) How can economic, social and political factors affect farming systems? [15]

2 (a) Discuss the characteristics of the informal sector of manufacturing and services. [10]

(b) Describe and explain the influence of three factors (e.g. labour) on the location of manufacturing and related service industry. [15]

Exam ready

12 Environmental management

12.1 Sustainable energy supplies

Non-renewable sources of energy are the fossil fuels and nuclear fuel. These are finite so that as they are used up the supply that remains is reduced. **Renewable energy** can be used over and over again. It includes hydroelectric, biomass, wind, solar, geothermal, tidal and wave power. At present, non-renewable resources dominate global energy. The challenge is to transform the global **energy mix** to achieve a better balance between renewable and non-renewable sources of energy.

> **Renewable energy** refers to sources of energy such as solar and wind power, which are not depleted as they are used.
>
> The **energy mix** is the relative contribution of different energy sources to a country's energy production/consumption.

Factors affecting the demand for, and supply of, energy

Revised

- Demand is primarily governed by the size of a country's population and its level of economic development.
- Growth in energy demand is particularly rapid in newly industrialised countries.
- A country's energy policy can impact significantly on demand if it focuses on efficiency and sustainability.
- High levels of pollution due to energy consumption can be a strong stimulus to developing a cleaner energy policy.

Global variations in energy supply occur for a number of reasons. For example:

Physical
Deposits of fossil fuels are only found in a limited number of locations.

Economic
In poor countries foreign direct investment is often essential for the development of energy resources.

Political
International agreements such as the Kyoto Protocol can have a considerable influence on the energy decisions of individual countries.

Trends in production and consumption

Revised

The fossil fuels dominate the global energy situation. Their relative contributions are (2008): oil – 34.8%, coal – 29.3%, natural gas – 24.1%. In contrast, hydroelectricity accounted for 6.4% and nuclear energy 5.5% of global energy. Figure 12.1 shows the regional pattern of energy consumption for 2008.

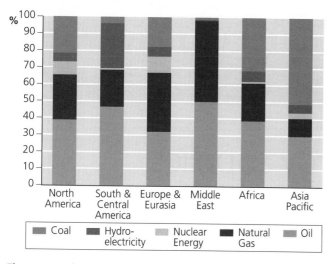

Figure 12.1 The regional pattern of energy consumption, 2008

- Deindustrialisation and increasing energy efficiency in MEDCs in general has resulted in a relatively modest increase in demand compared with newly industrialised countries (NICs). In some MEDCs the demand for energy has actually fallen.
- It is the NICs that are increasing their energy demand by the fastest rate. China alone has accounted for one-third of the growth in global oil demand since 2000.
- Most LEDCs struggle to fund their energy requirements. There is a strong positive correlation between GNP per capita and energy use. Around the world 2 billion people lack access to household electricity.

Oil

From less than 60 million barrels daily in the early 1980s global demand for oil rose steeply to 84.5 million barrels a day in 2008. The largest increase has been in the Asia-Pacific region, which now accounts for 30.1% of consumption. This region now uses more oil than North America (27.4%). In contrast, Africa consumes only 3.4% of global oil. The pattern of regional production is markedly different from that of consumption. In 2008, the Middle East accounted for 31.9% of production, followed by Europe and Eurasia (21.7%).

In 2008, the Middle East accounted for almost 60% % of global proved reserves. While the **reserves-to-production (R/P) ratio** is almost 79 years in the Middle East it is only 14.8 years in North America and 14.5 years in Asia-Pacific.

When will global peak oil production occur?

- The International Energy Agency expects **peak oil production** somewhere between 2013 and 2037.
- The Association for the Study of Peak Oil and Gas (ASPO) predicted that the peak of global oil production would come as early as 2011.

Other sources of energy

Between 1998 and 2008 global oil production increased by 11%. Over the same period this compared with:

- a rise of 35% in natural gas production. Natural gas production is dominated by the Russian Federation and the USA, accounting for 19.6% and 19.3% of the global total respectively.
- a 49% increase in coal production. China alone mines 42.5% of the world total. The next largest producing country is the USA (18.0%).

> **Typical mistake**
>
> Students sometimes confuse areas of production and consumption. For some energy sources such as coal the figures are very similar, but for oil there is a very significant difference. The ease with which a type of energy can be transported is the major factor here.

> The **reserves-to-production ratio** is the reserves remaining at the end of any year divided by the production in that year. The result is the length of time that those remaining reserves would last if production were to continue at that level.
>
> **Peak oil production** is the year in which the world or an individual oil producing country reaches its highest level of production, with production declining thereafter.

- a 13% increase in nuclear energy. With 103 operating reactors the USA leads the world in the use of nuclear electricity. This amounts to 31% of the world's total. The next major consumers of nuclear energy after the USA are France and Japan.
- a 22% rise in hydroelectricity. The 'big four' HEP nations of China, Canada, Brazil and the USA account for almost 50% of the global total.

Extending the 'life' of fossil fuels

There are a number of technologies that can improve the use and prolong the life of fossil fuels. These include coal gasification, clean coal technologies and the extraction of unconventional natural gas. Such techniques may be very important in buying time for more renewable energy to come on-line.

Renewable energy   Revised

Hydroelectricity dominates renewable energy production. The newer sources of renewable energy that make the largest contribution to global energy supply are wind power and biofuels.

Hydroelectric power

Most of the best HEP locations are already in use so the scope for more large-scale development is limited. However, in many countries there is scope for small-scale HEP plants to supply local communities.

Although HEP is generally seen as a clean form of energy it is not without its problems:

- Large dams and power plants can have a huge negative visual impact on the environment.
- The obstruction of the river affects aquatic life.
- Deterioration in water quality is common.
- Large areas of land may need to be flooded to form the reservoir behind the dam.
- Submerging large forests without prior clearance can release significant quantities of methane – a greenhouse gas.

Wind power

The worldwide capacity of wind energy is approaching 100,000 MW. Global wind energy is dominated by a relatively small number of countries. Germany, the USA and Spain together account for almost 58% of the world total. Wind energy has reached the 'take-off' stage, both as a source of energy and a manufacturing industry.

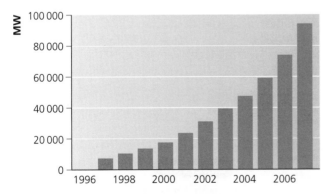

Figure 12.2 Wind energy – global cumulative installed capacity, 1996–2007

Apart from establishing new wind energy sites, **repowering** is also beginning to play an important role. This means replacing first-generation wind turbines with modern multi-megawatt turbines, which give a much better performance.

As wind turbines have been erected in more areas of more countries, the opposition to this form of renewable energy has increased:

- People are concerned that huge turbines located nearby could blight their homes and have a significant impact on property values.
- There are concerns about the hum of turbines disturbing both people and wildlife.
- Skylines in scenically beautiful areas might be spoiled forever.
- Turbines can kill birds. Migratory flocks tend to follow strong winds but wind companies argue that they steer clear of migratory routes.

Biofuels

Biofuels are fossil fuel substitutes that can be made from a range of agri-crop materials including oilseeds, wheat and sugar. They can be blended with petrol and diesel. Increasing amounts of cropland have been used to produce biofuels. Initially, environmental groups such as Friends of the Earth and Greenpeace were very much in favour of biofuels, but as damaging environmental consequences became clear such groups were the first to demand a rethink of this energy strategy.

Geothermal electricity

Geothermal energy is the natural heat found in the Earth's crust in the form of steam, hot water and hot rock. Rainwater can percolate several kilometres in permeable rocks where it is heated due to the Earth's **geothermal gradient**. This source of energy can be used to produce electricity or the hot water can be used directly for industry, agriculture, bathing and cleansing. For example in Iceland, hot springs supply water at 86°C to 95% of the buildings in and around Reykjavik.

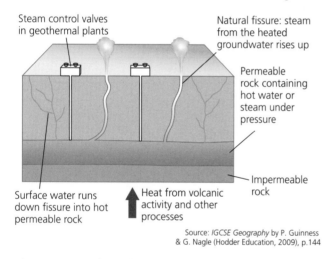

Steam control valves in geothermal plants

Natural fissure: steam from the heated groundwater rises up

Permeable rock containing hot water or steam under pressure

Impermeable rock

Surface water runs down fissure into hot permeable rock

Heat from volcanic activity and other processes

Source: *IGCSE Geography* by P. Guinness & G. Nagle (Hodder Education, 2009), p.144

Figure 12.3 Geothermal power

The USA is the world leader in geothermal electricity. However, total production accounts for just 0.37% of the electricity used in the USA. Other leading geothermal electricity-using countries are the Philippines, Italy, Mexico, Indonesia, Japan, New Zealand and Iceland.

Solar power

From a relatively small base the installed capacity of solar electricity is growing rapidly. Experts say that solar power has huge potential for technological improvement, which could make it a major source of global electricity in years to come. Spain, Germany, Japan and the USA currently lead the global market for solar power.

Solar power is currently produced in two ways:

- Photovoltaic systems – these are solar panels that convert sunlight directly into electricity.
- Concentrating solar power (CSP) systems use mirrors or lenses and tracking systems to focus a large area of sunlight into a small beam. This concentrated light is then used as a heat source for a conventional thermal power plant.

> **Expert tip**
>
> Solar power is generally taken to mean the production of solar electricity, as distinct from solar hot water systems, which are commonplace in many countries with a large number of days with sunshine each year.

Tidal power

Although tidal power is currently in its infancy, a study by the Electric Power Research Institute estimated that as much as 10% of US electricity could eventually be supplied by tidal energy. This potential could be equalled in the UK and surpassed in Canada. The 240 MW Rance facility in northwest France is the only utility-scale tidal power system in the world. The greatest potential for tidal power is in Canada's Bay of Fundy in Nova Scotia.

Fuelwood in developing countries

In developing countries about 2.5 billion people rely on fuelwood (including charcoal) and animal dung for cooking. Fuelwood accounts for just over half of global wood production. It provides much of the energy needs for Sub-Saharan Africa and is also the most important use of wood in Asia. In developing countries the concept of the 'energy ladder' is important. Here, a transition from fuelwood and animal dung to 'higher-level' sources of energy occurs as part of the process of economic development.

The environmental impact of energy   Revised ☐

The environmental impact of fossil fuels has been the subject of much debate over a long period of time, as have the significant concerns over nuclear energy. Even in the renewable sector the impact of large hydroelectric schemes has drawn considerable criticism along with increasing concerns over large wind farms. No energy project of a significant scale is without its disadvantages.

The Niger Delta

No energy production location has suffered more environmental damage than the Niger Delta, which contains over 75% of Africa's remaining mangrove. A 2006 report estimated that up to 1.5 million tonnes of oil have been spilt in the delta over the past 50 years. The report compiled by WWF says that the Delta is one of the five most polluted spots on Earth. Pollution is destroying the livelihoods of many of the 20 million people who live in the area. The pollution damages crops and fishing grounds and is a major contributor to the upsurge in violence in the region. People in the Delta are dissatisfied with bearing the considerable costs of the oil industry but seeing very little in terms of the benefits.

The flaring (burning) of unwanted natural gas found with the oil is a major regional and global environmental problem. The gas found here is not useful because there is no gas pipeline infrastructure to take it to consumer markets. Gas flaring in the Niger Delta is the world's single largest source of greenhouse gas emissions. The federal environmental protection agency has only existed since 1988 and **environmental impact assessments** were not compulsory until 1992.

Oil sands in Canada and Venezuela

Huge **oil sand** (tar sand) deposits in Alberta, Canada and Venezuela could be critical over the next 50 years as the world's production of conventional oil falls. However, there are serious environmental concerns about the development of tar sands:

- It takes two tonnes of mined sand to produce one barrel of synthetic crude, leaving lots of waste sand.
- It takes about three times as much energy to produce a barrel of Alberta oil-sands crude as it does a conventional barrel of oil. Thus, oil sands are large sources of greenhouse gas emissions.

An **environmental impact assessment** is a document required by law detailing all the impacts on the environment of a project above a certain size.

Oil sands, also known as tar sands or extra heavy oil, are naturally occurring mixtures of sand or clay, water and an extremely dense and viscous form of petroleum called bitumen.

Energy pathways crossing difficult environments

As energy companies have had to search further afield for new sources of oil, new **energy pathways** have had to be constructed. Some major oil and gas pipelines cross some of the world's most inhospitable terrain. The Trans-Alaska Pipeline crosses three mountain ranges and several large rivers. Much of the pipeline is above ground to avoid the permafrost. Engineers fly over the pipeline every day by helicopter to check for leaks and other problems. Incidents such as subsidence have closed the pipeline for short periods.

> **Energy pathways** are supply routes between energy producers and consumers, which may be pipelines, shipping routes or electricity cables.

Now test yourself

Tested

1 Define 'energy mix'.
2 List three technologies that can extend the life of fossil fuels.
3 What are the main environmental concerns about the development of hydroelectricity?
4 How important is fuelwood in LEDCs?

Answers on p.221

12.2 The management of energy supply

China overtook the USA in total energy usage in 2009. The demand for energy in China continues to increase significantly as the country expands its industrial base. In 2008, China's energy consumption breakdown by traditional energy sources was:

- coal – 70.2%
- oil – 18.7%
- hydroelectricity – 6.6%
- natural gas – 3.6%
- nuclear energy – 0.75%

> **Expert tip**
>
> This is a case study about the way in which the Chinese government has managed energy supply. Make sure you highlight the main decisions that the government has taken to ensure China's energy security.

China's energy strategy

Revised

China's energy policy has evolved over time. As the economy expanded rapidly in the 1980s and 1990s much emphasis was placed on China's main energy resource, coal (Figure 12.4). China was also an exporter of oil until the early 1990s; it is now a very significant importer. Chinese investment in energy resources abroad has risen rapidly. Long-term energy security is viewed as essential if the country is to maintain the pace of its industrial revolution.

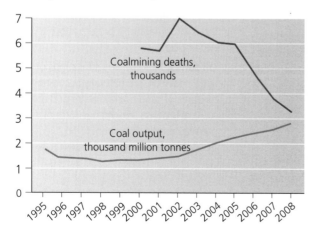

Figure 12.4 China's coal output, 1995–2008

In recent years China has tried to take a more balanced approach to energy supply and at the same time reduce its environmental impact through:

- energy conservation
- placing a strong emphasis on domestic resources
- diversified energy development
- environmental protection
- mutually beneficial international cooperation

The development of clean coal technology is an important aspect of China's energy policy. China has begun constructing clean coal plants at a rate of one a month and gradually retiring older, more polluting power plants. China has recently built a small experimental facility near Beijing to remove carbon dioxide from power station emissions and use it to provide carbonation for beverages.

The further development of nuclear and hydropower is another important strand of Chinese policy. The country also aims to stabilise and increase the production of oil while augmenting that of natural gas and improving the national oil and gas network.

China's strategic petroleum reserve
Priority was also given to building up the national oil reserve. In 2007 China announced an expansion of its crude reserves into a two-part system. Chinese reserves would consist of a government-controlled strategic reserve complemented by mandated commercial reserves. The government-controlled reserves are being completed in three phases. This will protect China to a certain extent from fluctuations in the global oil price, which can arise for a variety of reasons.

Renewable energy policy
Total renewable energy capacity in China reached 226 GW in 2009. This included:

- 197 GW of hydroelectricity
- 25.8 GW of wind energy
- 3.2 GW of biomass
- 0.4 GW of grid-connected solar PV

Renewable energy contributed more than one quarter of China's total installed energy capacity, with hydroelectricity by far the largest contributor. China's wind power capacity grew 30-fold between 2005 and 2009 to become the second largest in the world behind the USA. China's wind turbine manufacturing industry is now the largest in the world. It is now also the largest manufacturer of solar PV. China's current draft energy plan for 2020 sets targets of:

- 300 GW of hydroelectricity
- 150 GW of wind energy
- 30 GW of biomass
- 20 GW of solar PV

This would amount to almost one-third of China's planned power capacity of 1600 GW by 2020.

The Three Gorges Dam
The Three Gorges Dam across the Yangtze River includes the world's largest electricity generating plant of any kind. This is a major part of China's policy in reducing its reliance on coal. The dam is over 2 km long and 100 m high. The lake impounded behind it is over 600 km long. All of the originally planned components were completed in late 2008. Currently there are 32 main generators with a capacity of 700 MW each. Six additional generators in the underground power plant are being installed and should become fully operational in 2011. When totally complete the generating capacity of the dam will be a massive 22,500 MW. The Dam supplies Shanghai and Chongqing in particular with electricity. This is a multipurpose scheme that also increases the

river's navigation capacity and reduces the potential for floods downstream. However, there was considerable opposition to the dam for a number of reasons.

Now test yourself

5 When did China overtake the USA in total energy usage?
6 What are the main principles of China's current energy policy?
7 What name is given to China's stock of oil kept aside in case of an emergency?
8 Why has the Three Gorges Dam been so important to energy development in China?

Answers on p.221

12.3 Environmental degradation

Pollution: land, air and water — Revised

Pollution is the dominant factor in the **environmental degradation** of land, air and water. The most serious polluters are the large-scale processing industries (Table 12.1). Pollution is the major **externality** of industrial and urban areas. It is at its most intense at the focus of pollution-causing activities, declining with distance from such concentrations (Figure 12.5).

Pollution is contamination of the environment. It can take many forms – air, water, soil, noise, visual and others.

Environmental degradation is the deterioration of the environment through depletion of resources such as air, water and soil.

Externalities are the side effects (positive and negative) of an economic activity that are experienced beyond its site.

Table 12.1 Major sources and health and environmental effects of air pollutants

Pollutant	Major sources	Health effects	Environmental effects
Sulfur dioxide (SO_2)	Industry	Respiratory and cardiovascular illnesses	Precursor to acid rain, which damages lakes, rivers and trees; damage to cultural relics
Nitrous oxides (NOx)	Vehicles, industry	Respiratory and cardiovascular illnesses	Nitrogen deposition, leading to over-fertilisation and eutrophication
Particulate matter	Vehicles, industry	Particulates penetrate deep into the lungs and can enter the bloodstream	Reduced visibilty
Carbon monoxide (CO)	Vehicles	Headaches and fatigue, especially in people with weak cardiovascular health	–
Lead (Pb)	Vehicles (burning leaded petrol)	Accumulates in bloodstream over time; damages nervous system	Kills wildlife
Ozone (O_3)	Formed from a reaction of nitrous oxides and VOCs	Respiratory illnesses	Reduced crop production and forest growth; smog precursor
Volatile organic compounds (VOCs)	Vehicles, industrial processes	Eye and skin irritation; nausea, headaches; carcinogenic	Smog precursor

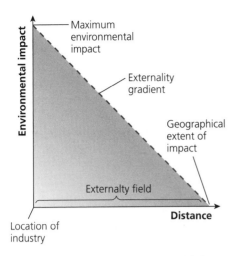

Figure 12.5 Externality gradient and field

Pollution control in the developed world

Levels of pollution have generally declined in the developed world because of:

- increasingly strict environmental legislation
- industry spending increasing amounts on research and development to reduce pollution
- the relocation of the most polluting activities to the emerging market economies

Thus the expectation is that after a certain stage of economic development in a country the level of pollution and the degradation it causes will decline.

It is important to consider the different impacts of **incidental pollution** and **sustained pollution**. The former is mainly linked to major accidents caused by technological failures and human error. Results of the latter include ozone depletion and global warming. Major examples of incidental pollution such as Chernobyl and Bhopal can have extremely long-lasting consequences. It is usually the poorest people in a society who are exposed to the risks from both incidental and sustained pollution.

> **Incidental pollution** is a one-off pollution incident.
>
> **Sustained pollution** is longer-term pollution.

Sustained pollution: ozone depletion and skin cancer

The ozone layer in the stratosphere prevents most harmful UV radiation from passing through the atmosphere. However, CFCs and other ozone-depleting substances have caused an estimated decline of about 4% a decade in the ozone layer since the late 1970s. This has allowed more UV radiation to reach the ground, leading to more cases of skin cancer, cataracts and other health and environmental problems.

Widespread global concern resulted in the Montreal Protocol banning the production of CFCs and related ozone depleting chemicals.

Sustained pollution, such as that caused by ozone-depleting substances, usually takes much longer to have a substantial impact on human populations than incidental pollution, but it is likely to affect many more people in the long term.

Water: demand, supply and quality

Revised

For about 80 countries, with 40% of the world's population, lack of water is a constant threat. And the situation is getting worse, with demand for water doubling every 20 years. In the poorest nations it is not just a question of lack of water; the paltry supplies available are often polluted. Securing access to clean water is a vital aspect of development. Out of the 2.2 million unsafe drinking water deaths in 2004, 90% were children under the age of 5. It has been estimated that 2.3 billion people live in **water-stressed areas** with 1.7 billion resident in **water-scarce areas.**

Water utilisation at the regional scale

There is a great imbalance in water demand and supply. Over 60% of the world's population live in areas receiving only 25% of global annual precipitation. Figure 12.6 shows how global water use has increased. In the developing world agriculture accounts for over 80% of total water use, with industry using more of the remainder than domestic allocation. In the developed world, agriculture accounts for slightly more than 40% of total water use. This is lower than the amount allocated to industry. As in the developing world domestic use is in third place.

> **Water-stressed areas** are where water supply is below 1700 m³ per person per year.
>
> **Water-scarce areas** are where water supply falls below 1000 m³ per person a year.

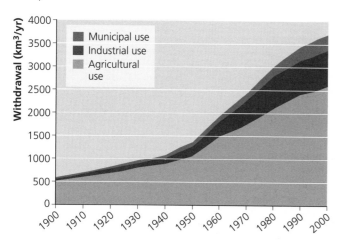

Figure 12.6 Global water use (agriculture, industry and domestic), 1900–2000

The environmental and human factors affecting water scarcity

Water scarcity is to do with the availability of potable water. **Physical water scarcity** is when physical access to water is limited. This is when demand outstrips a region's ability to provide the water needed by the population. It is the arid and semi-arid regions of the world that are most associated with physical water scarcity.

Economic water scarcity exists when a population does not have the necessary monetary means to utilise an adequate source of water. The unequal distribution of resources is central to economic water scarcity, where the problem is lack of investment. Much of Sub-Saharan Africa is affected by this type of water scarcity.

Scientists expect water scarcity to become more severe largely because:

● the world's population continues to increase significantly
● increasing affluence is inflating per capita demand for water
● of the increasing demands of biofuel production (biofuel crops are heavy users of water)
● climate change is increasing aridity and reducing supply in many regions
● many water sources are threatened by various forms of pollution

The degradation of rural environments ⎯⎯⎯ Revised ☐

Rural environments supply people with most of their food and contain the vast majority of the world's forested land. However, rural areas all around the world have been degraded at a rapid rate due to a range of factors, including overpopulation, poor agricultural practices and deforestation (Table 12.2). Agro-industrialisation has had a massive impact on rural ecosystems.

There is a growing realisation that the modes of production, processing, distribution and consumption that prevail, because in the short-to-medium term they are the most profitable, are not necessarily the most healthy or the most environmentally sustainable. In many parts of the world agro-industrialisation is having a devastating impact on the environment, causing:

- deforestation
- land degradation and desertification
- salinisation and contamination of water supplies
- air pollution
- increasing concerns about the health of long-term farm workers
- landscape change
- declines in biodiversity

Table 12.2 Five root causes of unsustainable practices

Cause	Explanation
Policy failure	Leading among the causes of unsustainable agriculture are inadequate or inappropriate policies, including pricing, subsidy and tax policies that have encouraged the excessive, and often uneconomic, use of inputs such as fertilisers and pesticides, and the overexploitation of land. They may also include policies favouring farming systems that are inappropriate both to the circumstances of the farming community and to available resources.
Rural inequalities	Rural people often know best how to conserve their environment, but they may need to overexploit resources in order to survive. Meanwhile, commercial exploitation by large landowners and companies often causes environmental degradation in pursuit of higher profits.
Resource imbalances	Almost all of the future growth in the world's population will be in LEDCs, and the biggest increases will be in the poorest countries of all – those least equipped to meet their own needs or invest in the future.
Unsustainable technologies	New technologies have boosted agricultural production worldwide, but some have had harmful side-effects, which must be contained and reversed, such as resistance of insects to pesticides, land degradation through wind or water erosion, nutrient depletion, poor irrigation management and the loss of biodiversity.
Trade relations	As the value of raw materials exported by LEDCs has fallen, their governments have sought to boost income by expansion of crop production and timber sales, which has damaged the environment.

Agro-industrialisation is characterised by large areas of monoculture, which leaves crops more vulnerable to disease due to the depletion of natural systems of pest control. Monoculture results in reliance on pesticides, which leads to a downward environmental cycle. Large-scale farming has been expanding geographically into a number of fragile environments, particularly into areas of rainforest. The global cattle population is currently around 1.5 billion. The balance between livestock and grass is sustainable at present, but as the demand for meat increases, the pressures that cattle make on the land may well soon exceed supply.

Poor households are often compelled to degrade environmental resources. They can also suffer significantly from the actions of large-scale rural operations and may be pushed onto more marginal lands by logging, ranching or mining operations.

Expert tip

Remember that urban areas can affect the environmental degradation of their rural surroundings in a number of ways. For example, untreated urban wastewater is a major pollutant of rivers and landfill sites for urban waste may have adverse impacts.

The degradation of urban environments

Revised

The degradation of urban environments occurs mainly through urbanisation, industrial development and inadequate infrastructure. Figure 12.7 shows how urban environmental problems can impact at the household, community and city scales. The urban poor are particularly affected by poor environmental services such as sub-standard housing, lack of sanitation and other aspects of urban poverty.

Amenity loss	Traffic congestion	**City**	Loss of heritage and historical buildings	Reduced property and building values
Accidents and disasters	Polluted land	**Community**	Inappropriate and inadequate technology use	Inadequate tax/financial revenues
Flooding and surface drainage	Garbage dumping	**Household** Household health, garbage generation, air/water/noise pollution, spread of diseases	Lack of understanding of environmental problems	Lack of, and inappropriate, laws and legislation
Toxic and hazardous wastes/dumps				High living densities
	Flooding	Noise pollution	Natural disasters	
Loss of agricultural land and desertification	Air pollution	Water pollution	Inadequate supply and transmission loss of electricity	Misguided urban, government and management practices

Figure 12.7 The scales of urban environmental problems

A recent study concluded that 16 of the world's 20 most polluted cities are in China. 75% of urban residents breathe in polluted air. In 1998 the World Resources Institute declared Lanzhou the world's most polluted city. It is a major industrial centre, burning large quantities of coal every day. The city is surrounded by hills, which hinder the dispersal of pollution. Since then the city has addressed its environmental problems by:

- attempting to close some heavy industries
- relocating some industries from inner-city to edge-of-city locations
- restricting emissions from factories
- investing in supplying natural gas and cleaner coal
- restricting traffic
- planting trees on surrounding hillsides to reduce dust storms

In July 2009, the Beijing municipal government launched a strategy for the improved disposal of the 18,410 tonnes of domestic refuse generated by the city every day. These measures include building more environmentally friendly disposal sites, improving incineration technologies, and enforcing household waste separation and recycling. The last Saturday of every month has been designated as a 'recyclable resources collecting day'. However, people in China are not yet as 'recycling aware' as populations in richer countries. Much needs to be done in terms of environmental education to improve this situation.

Beijing will run out of space for landfills in just 4 years. More than a third of Chinese cities are facing a similar crisis. At present, only 10% of waste is incinerated. However, significantly increasing incineration requires substantial investment and there is always considerable opposition from people living close to the sites selected for new incinerators. The main concern is the significant amounts of heavy metals and dioxins released into the atmosphere by incinerators.

Constraints on improving degraded environments
Revised

There are numerous constraints on improving the quality of degraded environments:

- In many parts of the developing world population growth continues at a high rate.
- High rates of rural–urban migration can lead to rapidly deteriorating environmental conditions in large urban areas.
- Environmental hazards, often made worse by climate change, present an increasing challenge in some world regions.

- Poor knowledge about the environmental impact of human actions is a significant factor in many locations.
- Poor management at both central and local government levels may result in problems that can be at least partially rectified, not being addressed.
- Many degraded environments require substantial investment to bring in realistic solutions.
- Civil war has put back development by decades in some countries.
- Corruption and crime can also reduce the effectiveness of schemes to reduce environmental degradation.

The protection of environments at risk
Revised

At the most extreme, human activity and access can be totally banned, such as in Wilderness Areas, or extremely limited, as is usually the case in National Parks. However, in many areas it is usually necessary to sustain significant populations and rates of economic activity, particularly in developing countries. In these cases various types of sustainable development policies need to be implemented. Individual environments can be assessed in terms of:

- needs – what needs to be done to reduce environmental degradation as far as possible without destroying the livelihoods of the resident population?
- measures – what are the policies and practices that can be implemented to achieve these aims at various time scales?
- outcomes – how successful have these policies been at different stages of their implementation? Have policies been modified to cope with initially unforeseen circumstances?

Now test yourself
Tested

9 Define environmental degradation.
10 Give **two** reasons why levels of pollution have generally declined in the developed world.
11 Distinguish between water-stressed areas and water-scarce areas.
12 How has agro-industrialisation degraded rural environments?

Answers on p.221

12.4 The management of a degraded environment

Namibia is a poor and sparsely populated country in southwest Africa. (Figure 12.8) Environmental degradation and sustainability are significant issues in Namibia's marginal landscapes. The government is attempting to tackle these issues and reduce poverty at the same time. The causes of degradation have been mainly uncontrolled exploitation by a low-income population and lack of management at all levels of government in earlier years.

Namibia's Communal Conservancy Program is regarded as a successful model of community-based natural resource management. The program gives rural communities unprecedented management and use rights over wildlife, which have created new incentives for communities to protect this valuable resource and develop economic opportunities in tourism.

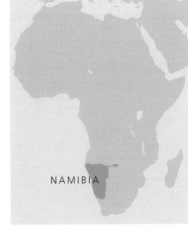

Figure 12.8 Location of Namibia

The conservancy program began in 1996. By 2007 it had expanded to 50 registered **communal conservancies**, which now cover nearly 11.9 million hectares. This encompasses over 14% of the area of the country, benefiting more than 230,000 rural dwellers. An obvious sign of success is the significant increase in the numbers of wildlife in the conservancies after decades of decline.

The conservancies benefit from a number of 'new' economic activities including:

● contracts with tourism companies
● selling hunting concessions
● managing campsites
● selling wildlife to game ranchers
● selling crafts

These activities are in addition to traditional farming practices, which were traditionally at the subsistence level. The diversification of economic activity made possible by the conservancy programme has increased employment opportunities and raised incomes. Support from, and cooperation between, a number of different institutions has been important to the development of the programme. Such institutions bring substantial experience and skills in helping conservancies to develop. Figure 12.9 shows the rapid expansion of the total land area under management of conservancies from 1998 to 2005.

> **Communal conservancies** are legally recognised common property resource management organisations in Namibia's communal lands.

> **Typical mistake**
>
> It is easy to think that managing degraded environments means a substantial reduction in economic activity, when often what is required is a significant change in economic activity. Such changes, if properly managed, can both improve the environment and improve incomes for the resident population.

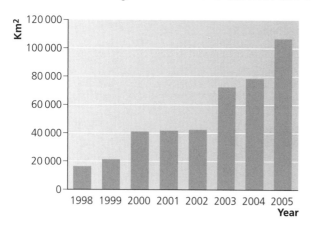

Figure 12.9 Communal conservancies – total land under management, 1998–2005

Although rural poverty remains significant in Namibia, the conservancy programme has resulted in substantial progress, with income rising year on year. Rising income from conservancies has made possible increasing investment in social development projects. This has made conservancies an increasingly important element in rural development.

Following the perceived success of community conservancies, the Namibian government has extended the concept to **community forests**. Establishing a community forest is similar to the process of forming a conservancy. This is a good example of the **scaling-up process** from one natural resource system to another.

> **Community forests** are forest areas where local communities can manage and utilise forest resources in a sustainable manner.
>
> The **scaling-up process** involves expanding effective programmes to reach larger numbers of people in a broader geographic area.

Now test yourself

13 What are communal conservancies?

14 When did Namibia's communal conservancy program begin?

15 By how much did the total land under communal conservancy management increase from 1998 to 2005?

16 Which new economic activities have helped increase incomes in communal conservancies?

Answers on p.221

Exam-style questions

1 **(a)** Examine the factors that can affect the demand for energy.   [10]

   (b) With reference to examples, discuss the environmental impact of energy production.   [15]

2 **(a)** What are the problems associated with the demand for, and supply of, water?   [10]

   (b) Discuss the factors responsible for the degradation of rural environments.   [15]

Exam ready

13 Global interdependence

13.1 Trade flows and trading patterns

Global trade

Revised

Trade refers to the exchange of goods and services for money. World trade now accounts for 25% of GDP, double its share in 1970. Goods and services purchased from other countries are termed **imports**. In contrast, goods and services sold to other countries are called **exports**. The difference between the two is known as the **balance of trade**. **Visible trade** involves items that have a physical existence and can actually be seen, such as oil and manufactured goods. **Invisible trade** is trade in services, which include travel and tourism, and business and financial services.

> **Trade deficit** is when the value of a country's imports exceeds the value of its exports.
>
> **Trade surplus** is when the value of a country's exports exceeds the value of its imports.

Global inequalities in trade flows

Europe, Asia and North America dominate global trade (Table 13.1). Germany was the largest exporter of merchandise in 2008 with 9.1% of the global share, followed by China and the USA. The top 10 countries accounted for 50.7% of world exports. However, the USA dominates imports by a huge margin – over 13% of the world total, followed by Germany and China. The share of developing economies in world merchandise trade set new records in 2008. The emergence of different generations of newly industrialised countries since the 1960s has radically altered the trade pattern that existed in the previous period.

Table 13.1 World merchandise trade by region, 2008

Region	Exports ($ billion)	Imports ($ billion)
World	15775	16120
North America	2049	2909
South and Central America	602	595
Europe	6459	6833
Commonwealth of Independent States	703	493
Africa	561	466
Middle East	1047	575
Asia	4355	4247

Source: WTO

In recent decades trade in commercial services has increased considerably. However, in terms of total value it is still less than a quarter of that of merchandise trade.

Factors affecting global trade

Resource endowment

The Middle East countries dominate the export of oil because of their large oil reserves. Countries endowed with other raw materials such as food products, timber, minerals and fish also figure prominently in world trade statistics. In the developed world the wealth of countries such as Canada and Australia has been built to a considerable extent on the export of raw materials in demand on the world market. Raw-material-rich developing countries such as Brazil and South Africa have been trying to follow a similar path. In both cases, wealth from raw materials has been used for economic diversification to produce more broadly based economies.

Locational advantage

The location of market demand influences trade patterns. It is advantageous for an exporting country to be close to the markets for its products; for example, it reduces transport costs. Manufacturing industry in Canada benefits from the proximity of the huge US market. Some countries and cities are strategically located along important trade routes, giving them significant advantages in international trade. For example, Singapore, at the southern tip of the Malay peninsula is situated at a strategic location along the main trade route between the Indian and Pacific Oceans.

Investment

Investment in a country is the key to increasing trade. Some developing countries such as India and Mexico have increased their trade substantially. These countries have attracted high levels of foreign direct investment. However, two billion people live in countries that have become less, rather than more, globalised as trade has fallen in relation to national income. This group includes many African countries.

Historical factors

Historical relationships, often based on colonial ties, remain an important factor in global trade patterns. For example, the UK still maintains significant trading links with Commonwealth countries. Other European countries such as France and Spain also established colonial networks overseas and have maintained such ties to varying degrees. Colonial expansion heralded a trading relationship dictated by the European countries mainly for their own benefit. The colonies played a subordinate role, which brought them only very limited benefits at the expense of distortion of their economies. The historical legacy of this **trade dependency** is one of the reasons why poorer tropical countries have such a limited share of world trade according to development economists.

The terms of trade

The most vital element in the trade of any country is the terms on which it takes place. If countries rely on the export of commodities that are low in price and need to import items that are relatively high in price they need to export in large quantities to be able to afford a relatively low volume of imports. Many poor nations are **primary product dependent**. The world market price of primary products is in general very low compared with manufactured goods and services. The **terms of trade** for many developing countries are worse now than they were two decades ago. Because the terms of trade are generally disadvantageous to the poor countries of the South, many developing countries have very high trade deficits (Figure 13.1).

> **Trade dependency** is when a developing country is so reliant on its advanced trading partner(s) that any changes in their economic policy or economic condition could have a severe effect on the developing country's economy.
>
> **Primary product dependence** is when countries rely on one or a small number of primary products for the bulk of their export earnings.
>
> The **terms of trade** refer to the price of a country's exports relative to the price of its imports, and the changes that take place over time.

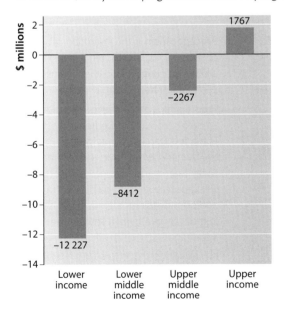

Figure 13.1 Trade balance by income group

Changes in the global market

The rapid growth of newly industrialised countries has brought about major changes in the economic balance of power. The substantial growth rates of the BRIC countries (Brazil, Russia, India, China) in particular are very much a threat to the established economic order. These four countries, along with other high-growth nations outside of the established core group of nations, are known as **emerging markets**. While the developed world (the core) grew by an average of 2.1% a year in the first decade of the century, the emerging markets expanded by 4.2%.

In 1990 the developed world controlled about 64% of the global economy; this fell to 52% by 2009 – one of the most rapid economic changes in history. The West no longer dominates the world's savings and as a result no longer dominates global investment and finance.

Trade agreements

A **trade bloc** is a group of countries that share trade agreements between each other. Since the Second World War there have been many examples of groups of countries joining together to stimulate trade between themselves and to obtain other benefits from economic cooperation. Regional trade agreements have proliferated in the last two decades. In 1990 there were fewer than 25; by 1998 there were more than 90. The most notable of these are the European Union, NAFTA in North America, ASEAN in Asia, and Mercosur in Latin America.

> **Expert tip**
>
> Trade blocs have varying levels of economic integration from free trade areas such as NAFTA to the much higher level of integration of an economic union, as illustrated by the EU.

Trade and development   Revised ☐

There is a strong relationship between trade and economic development. In general, countries which have a high level of trade are richer than those which have lower levels of trade. Countries which can produce goods and services in demand elsewhere in the world will benefit from strong inflows of foreign currency and from the employment its industries provide. Foreign currency allows a country to purchase abroad goods and services it does not produce itself or produce in large enough quantities. An Oxfam report published in April 2002 stated that if Africa increased its share of world trade by just 1% it would earn an additional £49 billion a year – five times the amount it receives in aid.

The World Trade Organization

The World Trade Organization (WTO) was established in 1995. Unlike its predecessor, the loosely organised GATT, the WTO was set up as a permanent organisation with far greater powers to arbitrate trade disputes. Today average tariffs are only a tenth of what they were when GATT came into force and world trade has been increasing at a much faster rate than GDP. However, in some areas **protectionism** is still alive and well, particularly in clothing, textiles and agriculture. In principle, every nation has an equal vote in the WTO. In practice, the rich world shuts the poor world out in key negotiations. In recent years agreements have become more and more difficult to reach.

The WTO exists to promote **free trade**. Most countries in the world are members and most who are not want to join. The fundamental issue is: does free trade benefit all those concerned or is it a subtle way in which the rich nations exploit their poorer counterparts? Most critics of free trade accept that it does generate wealth but they deny that all countries benefit from it.

Critics of the WTO:

- say that the WTO should be paying more attention to the needs of poor countries, making it easier for them to gain tangible benefits from the global economic system
- ask why it is that MEDCs have been given decades to adjust their economies to imports of textiles and agricultural products from LEDCs, when the

> **Protectionism** is the institution of policies (tariffs, quotas, regulations) that protect a country's industries against competition from cheap imports.
>
> **Free trade** is a hypothetical situation whereby producers have free and unhindered access to markets everywhere. While trade is more free today than in the past, governments still impose significant barriers to trade and often subsidise their own industries in order to give them a competitive advantage.

latter are pressurised to open their borders immediately to MEDCs banks, telecommunications companies and other components of the service sector

Opposition to the WTO comes from a number of sources:

- many developing countries who feel that their concerns are largely ignored
- environmental groups concerned, for example, about a WTO ruling that failed to protect dolphins from tuna nets
- labour unions in some developed countries, notably the USA, concerned about (a) the threat to their members' jobs as traditional manufacturing filters down to LEDCs and (b) violation of 'workers' rights' in developing countries

---

Case study **The trade in tea**

Tea, like coffee, bananas and other raw materials, exemplifies the relatively small proportion of the final price of the product that goes to producers (Figure 13.2). The great majority of the money generated by the tea industry goes to the post-raw-material stages (processing, distributing and retailing), usually benefiting companies in MEDCs rather than the LEDC producers.

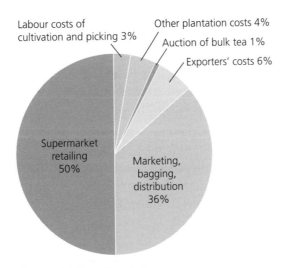

Figure 13.2 Tea value chain

The nature and role of fair trade

Many supermarkets and other large stores in Britain and other MEDCs now stock some fairly traded products. Most are agricultural products such as bananas and tea, but the market in non-food goods such as textiles and handicrafts is also increasing. The fair trade system operates as follows:

- Small-scale producers group together to form a cooperative with high social and environmental standards.
- These cooperatives deal directly with companies such as Tesco in MEDCs.
- MEDC companies pay significantly over the world market price for the products traded.
- The higher price achieved by the LEDC cooperatives provides both a better standard of living and some money to reinvest in their farms.

Advocates of the fair trade system argue that it is a model of how world trade can and should be organised to tackle global poverty. It began in the 1960s with Dutch consumers supporting Nicaraguan farmers. It is now a global market worth £315 million a year, involving over 400 MEDC companies and an estimated 500,000 small farmers and their families in the world's poorest countries.

Fair trade is a movement that aims to create direct, long-term trading links with producers in developing countries, and to ensure that they receive a guaranteed price for their products, on favourable financial terms.

Now test yourself

1 What is the difference between visible and invisible trade?
2 What are the main factors affecting global trade?
3 Briefly examine the link between trade and development.
4 Describe and explain the nature and role of fair trade.

Answers on p.221

Tested

13.2 Debt and aid and their management

Debt: causes, nature and problems

Revised

Many experts single out **debt** as the major problem for the world's poorer nations. The term debt generally refers to external debt (foreign debt). Many poor countries are currently paying back large amounts in debt repayments while at the same time struggling to provide basic services for their populations. An ever-increasing proportion of new debt is used to service interest payments on old debts. The **debt service ratio** of many poor countries is at a very high level. Critics say that developed countries should do more to help the poor countries through **debt relief** and by opening their markets to exports from developing countries. The total external debt of the poorest countries of the world (the 'low-income countries') was $375 billion in 2006.

How did the debt crisis come about?

The **debt crisis** began with the Arab-Israeli war of 1973–74, which resulted in a sharp increase in oil prices. Governments and individuals in the oil-producing countries invested the profits from oil sales in the banks of MEDCs.

Eager to profit from such a high level of investment, these banks offered relatively low-interest loans to poorer countries. They were encouraged to exploit raw materials and grow cash crops so that they could pay back their loans with profits made from exports. However, periods of recession in the 1980s and 1990s led to rising inflation and interest rates. At the same time crop surpluses resulted in a fall in prices.

As a result, the demand for exports from developing countries fell and export earnings declined significantly as a result. These factors, together with oil price increases, left many developing countries unable to pay the interest on their debts.

Loans can help countries to expand their economic activities and set up an upward spiral of development if used wisely. However, many of the loans that burden the world's poorest countries were given under dubious circumstances and at times at very high rates of interest. Many development economists also focus on the legacy of **colonialism**, arguing that the colonising powers left their former colonies with high and unfair levels of debt when they became independent.

In recent years much of the debt has been 'rescheduled' and new loans have been issued. However, new loans have frequently been granted only when poor countries agreed to very strict conditions under 'structural adjustment programmes', which have included:

- agreeing to free trade measures, which have opened up their markets to intense foreign competition
- severe cuts in spending on public services such as education and health
- the privatisation of public companies

However, despite the disadvantages that many countries have suffered over the medium and long-term from improper lending, debt is a vital component of the global economic system.

Figure 13.3 is a Christian Aid newspaper advertisement illustrating the plight of Haiti, one of the world's poorest countries, after the devastating earthquake of January 2010. Some countries need to put aside between 20% and 30% of their export earnings to meet their debt repayments, which can prove to be a crippling burden for nations with very low incomes.

Debt in this context refers to external debt (foreign debt), which is that part of the total debt in a country owed to creditors outside the country. The debtors can be the government, corporations or private households. The debt includes money owed to commercial banks, other governments, or international financial institutions such as the World Bank.

Debt service ratio is the ratio of debt service payments of a country to that country's export earnings.

Debt relief is cancellation of debts owed by developing nations to industrialised nations or institutions such as the World Bank, in order to allow the government to shift funds toward social development.

Typical mistake

Sometimes student's answers imply that only poor countries have significant debts. While LEDCs owe substantial amounts of money, MEDCs have borrowed much more. The USA owes more money to the rest of the world than any other country. The point here is that rich countries have huge assets against which they can borrow.

A **debt crisis** occurs if major debtors are unable or unwilling to pay the interest and redemption payments due on their debts, or if creditors are not confident they will do so.

Colonialism is the building and maintaining of colonies in one territory (or a number of territories) by people from another territory.

THE ONLY THING STILL STANDING IS THE DEBT.

The suffering in Haiti is extraordinary. Tens of thousands dead. Many more injured and homeless. The British public have donated over £38m to the DEC to get emergency aid to the people suffering on the ground. It's been a fantastic response.

But when the emergency response teams have done their work, the Haitians will still face an uphill battle. Haiti has debts of over $800m. No wonder 80% of the population lives in poverty. As part of our ongoing fight against the systems that keep people poor, Christian Aid is lobbying Chancellor Alistair Darling (the UK's representative at the IMF) to lead the world in making sure that all of Haiti's debt is cancelled. You can help us.

Sign the petition online to drop the debt.
Visit www.christianaid.org.uk

POVERTY

christian aid

A009667

Figure 13.3 Christian Aid Haiti campaign

Debt relief and aid

- Restructuring debt to developing countries began in a limited way in the 1950s.
- Attempts were made by creditor nations to tackle the 'Third World debt crisis' throughout the 1980s and 1990s. However, these efforts were viewed as limited in nature and the overall debt of poorer countries continued to rise.
- It was not until the mid-1990s that a more comprehensive global plan to tackle the debt of the poorest countries was formulated.

The Heavily Indebted Poor Countries (HIPC) initiative

The HIPC initiative was first established in 1996 by the IMF and the World Bank. Its aim was to provide a comprehensive approach to debt reduction for heavily indebted poor countries so that no poor country faced a debt burden it could not manage. To qualify for assistance countries have to pursue IMF and World Bank supported adjustment and reform programmes. According to a recent World Bank-IMF report debt relief provided under both initiatives has substantially alleviated debt burdens in recipient countries.

Debt relief is part of a much larger process, which includes international aid, designed to address the development needs of low-income countries. For debt reduction to have a meaningful impact on poverty, the additional funds made available need to be spent on programmes that are of real benefit to the poor. Before the HIPC initiative, eligible countries spent on average more on debt serving than on education and health combined. Now these countries have significantly increased their spending on education, health and other social services.

International aid

Aid is assistance in the form or grants or loans at below market rates. Most developing countries have been keen to accept foreign aid because of:

- the 'foreign exchange gap', whereby many developing countries lack the hard currency to pay for imports such as oil and machinery, which are vital to development
- the 'savings gap', where population pressures and other drains on expenditure prevent the accumulation of enough capital to invest in industry and infrastructure
- the 'technical gap', caused by a shortage of skills needed for development

> **International aid** is the giving of resources (money, food, goods, technology etc.) by one country or organisation to another, poorer country. The objective is to improve the economy and quality of life in the poorer country.

The different types of international aid

Figure 13.4 shows the different types of **international aid**. The basic division is between:

- official government aid
- voluntary aid, run by non-governmental organisations (NGOs) or charities such as Oxfam

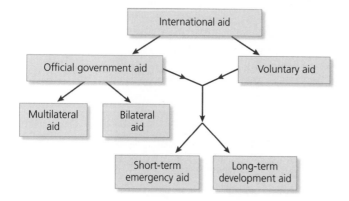

Figure 13.4 The different types of international aid

Official government aid can be divided into:

- bilateral aid, which is given directly from one country to another
- multilateral aid, which is provided by many countries and organised by an international body such as the UN

Aid supplied to poorer countries is of two types:
- short-term emergency aid, which is provided to help cope with unexpected disasters such as earthquakes
- long-term development aid, which is directed towards the continuous improvement in the quality of life in a poorer country

There is no doubt that many countries have benefited from international aid. All the countries that have developed into NICs from LEDCs have received international aid. But their development has also been down to other factors. It is difficult to be precise about the contribution of international aid to the development of each country. According to some left-wing economists aid does not do its intended job because:
- it often fails to reach the very poorest people and when it does the benefits are frequently short-lived
- a significant proportion of foreign aid is '**tied**' to the purchase of goods and services from the donor country
- the use of aid on large capital intensive projects may actually worsen the conditions for the poorest people
- aid can delay the introduction of reforms, for example the substitution of food aid for land reform
- international aid can create a culture of dependency, which can be difficult to break

Arguments put forward by the political right against aid are as follows:
- Aid encourages the growth of a larger-than-necessary public sector.
- The private sector is 'crowded out' by aid funds.
- Aid distorts the structure of prices and incentives.
- Aid is often wasted on grandiose projects of little or no benefit to the majority of the population.
- The West did not need aid to develop.

Many development economists argue there are two issues more important to development than aid: (a) changing the terms of trade so that developing nations get a fairer share of the benefits of world trade; (b) writing off the debts of the poorest countries. Figure 13.5 shows official development assistance (ODA) received by world region.

> **Tied aid** is foreign aid that must be spent in the country providing the aid (the donor country).

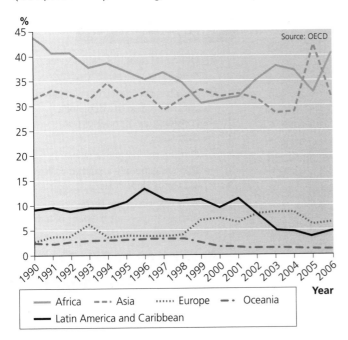

Figure 13.5 ODA by world region

The effectiveness of aid: top–down and bottom–up approaches

Over the years most debate about aid has focused on the amount of aid made available. However, in recent years the focus has shifted somewhat to the effectiveness of aid. This has involved increasing criticism of the traditional top–down approach to aid. The Hunger Project is one of a number of organisations that have adopted a radically different approach. It has worked in partnership with grassroots organisations in Africa, Asia and Latin America to develop effective bottom–up strategies. The key strands in this approach have been:

- mobilising grassroots people for self-reliant action
- intervening for gender equality
- strengthening local democracy

NGOs: leading sustainable development

NGOs have often been much better at directing aid towards sustainable development than government agencies. The selective nature of such aid has targeted the poorest communities using **appropriate technology** and involving local people in decision-making. For example, Water Aid, established in 1981, is dedicated exclusively to the provision of safe domestic water, sanitation and hygiene education to the world's poorest people. This combination maximises health benefits and promotes development (Figure 13.6). In the longer term, communities are able to plan and build infrastructure that enables them to cope better in times of hardship.

> **Appropriate technology** is aid supplied by a donor country whereby the level of technology and the skills required to service it are properly suited to the conditions in the receiving country.

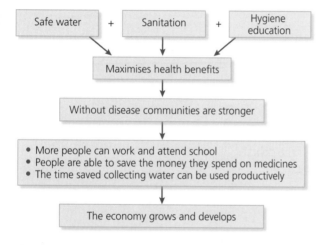

Figure 13.6 Water Aid's building blocks of development

> **Microcredit** involves tiny loans and financial services to help the poor – mostly women – start businesses and escape poverty.

Microcredit

The development of the Grameen Bank in Bangladesh illustrates the power of **microcredit** in the battle against poverty. The Grameen Foundation uses microfinance and innovative technology to fight global poverty and bring opportunities to the poorest people. The bank provides tiny loans and financial services to poor people to start their own businesses. Women are the beneficiaries of most of these loans. A typical loan might be used to buy a cow to sell milk to fellow villagers, or to purchase a piece of machinery that can be hired out to other people in the community. The concept has spread beyond Bangladesh to reach 3.6 million families in 25 countries.

Now test yourself

5 Define (a) debt, (b) debt service ratio.

6 How has the legacy of colonialism contributed to the debts of a considerable number of poor countries?

7 Outline the sequence of events that are generally accepted to have led to the debt crisis.

8 Why is appropriate technology such an important element of aid to poor countries?

Answers on pp.221–222

Tested

13.3 The development of international tourism

Reasons for, and trends in, the growth of tourism

Tourism has developed into a major global industry, which is still expanding rapidly. **International tourist arrivals** reached a record of almost 900 million in 2007 and **international travel receipts** totalled over $800 billion. Between 2000 and 2007 international tourist arrivals rose over 40%. A range of factors have been responsible for the growth of global tourism. Table 13.2 subdivides these factors into economic, social and political reasons and also includes factors that can reduce levels of tourism, at least in the short term.

> **Tourism** is travel away from the home environment (a) for leisure, recreation and holidays, (b) to visit friends and relatives, and (c) for business and professional reasons.
>
> **International tourist arrivals** are tourists travelling to a country that is not their place of residence, for more than one day but no longer than a year.
>
> **International tourism receipts** are money spent by visitors from abroad in a destination country.

Table 13.2 Factors affecting global tourism

Economic	Steadily rising incomes – tourism grows on average 1.3 times faster than GDP.
	The decreasing real costs (with inflation taken into account) of holidays.
	The widening range of destinations within the middle-income range.
	The heavy marketing of shorter foreign holidays aimed at those who have the time and disposable income to take an additional break.
	The expansion of budget airlines.
	'Air miles' and other retail reward schemes aimed at travel and tourism.
	'Globalisation' has increased business travel considerably.
	Periods of economic recession can reduce levels of tourism considerably.
Social	An increase in the average number of days of paid leave.
	An increasing desire to experience different cultures and landscapes.
	Raised expectations of international travel with increasing media coverage of holidays, travel and nature.
	High levels of international migration over the last decade or so means that more people have friends and relatives living abroad.
	More people are avoiding certain destinations for ethical reasons.
Political	Many governments have invested heavily to encourage tourism.
	Government backing for major international events such as the Olympic Games and the World Cup.
	The perceived greater likelihood of terrorist attacks in certain destinations.
	Government restrictions on inbound/outbound tourism.
	Calls by non-governmental organisations to boycott countries such as Burma.

Figure 13.7 shows the regional share of tourist arrivals in 2007. Although the developed regions of the world remain the largest tourism destinations, their dominance is reducing. For example, Europe and North America accounted for 69% of international arrivals in 2000, but by 2007 this had fallen to 62%. Between 2000 and 2007 the fastest rates of growth were in the Middle East, Asia and Africa. Many developing countries have become more open to foreign direct

investment in tourism compared with two or three decades ago. Tourism is one of the top five export categories for as many as 83% of countries and is the main source of foreign exchange for at least 38% of countries.

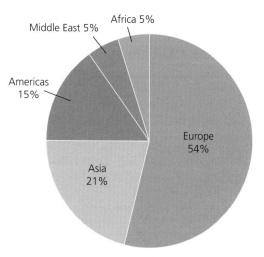

Figure 13.7 Share of tourist arrivals, 2007

However, tourism is vulnerable to 'external shocks'. Periods of economic recession, characterised by high unemployment, modest wage rises, and high interest rates, affect the demand for tourism in most parts of the world.

The social and cultural impact

Revised

Many communities in the developing world have suffered considerable adverse cultural changes, including:

- the loss of locally owned land
- the abandonment of traditional values and practices
- displacement of people to make way for tourist developments
- abuse of human rights by governments and companies in the quest to maximise profits
- alcoholism and drug abuse
- crime and prostitution, sometimes involving children
- visitor congestion at key locations hindering the movement of local people
- denying local people access to beaches to provide 'exclusivity' for visitors
- the loss of housing for local people as more visitors buy second homes in popular tourist areas

The attitudes towards tourism of host communities can change over time. An industry that is usually seen as very beneficial initially can eventually become the source of considerable irritation, particularly where there is a big clash of cultures. However, tourism can also have positive social and cultural impacts, such as:

- increasing the range of social facilities for local people
- helping develop foreign language skills in host communities

Expert tip

It is easy to fall into the trap of only seeing disadvantages in terms of the cultural impact of tourism because so much has been written about this topic. However, it is always important to consider the other side of the coin even if you can only come up with a few points.

The economic impact

Revised

The World Travel and Tourism Council has developed Travel and Tourism Satellite Accounting to show the full economic impact of tourism. By including all the direct and indirect economic implications of tourism it is clear that the industry has a much greater impact than most people think (Figure 13.8).

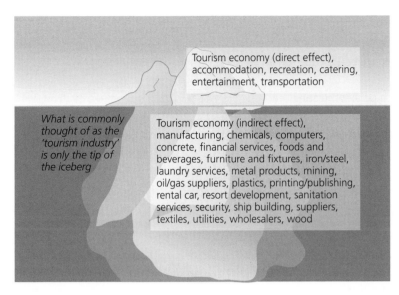

Figure 13.8 Satellite accounting 'iceberg'

Tourism undoubtedly brings valuable foreign currency to developing countries and a range of other obvious benefits but critics argue that its value is often overrated for the following reasons:

- **Economic leakages** (Figure 13.9) from developing to developed countries run at a rate of between 60% and 75%.
- Tourism is labour intensive, providing a range of jobs especially for women and young people. However, most local jobs created are low paid and seasonal.
- Money borrowed to invest in the necessary infrastructure for tourism increases the national debt.
- At some destinations tourists spend most of their money in their hotels with minimum benefit to the wider community.
- Tourism might not be the best use for local resources, which could in the future create a larger multiplier effect if used by a different economic sector.
- Locations can become over-dependent on tourism.

The tourist industry has a huge appetite for basic resources, which often impinges heavily on the needs of local people. In such situations tourist numbers may exceed the **carrying capacity** of a destination by placing too much of a burden on local resources.

Economic leakages comprise the part of the money a tourist pays for a foreign holiday that does not benefit the destination country because it goes elsewhere, such as payments to foreign owners of hotels.

The **carrying capacity** of a destination is the number of tourists a destination can take without placing too much pressure on local resources and infrastructure.

The **multiplier effect** occurs when a new or expanding economic activity in a region creates new employment and increases the amount of money circulating in that region. In turn, this attracts further economic development, creating more employment, services and wealth.

LEDC tourist destinations

Total money spent on tourism to this destination

Transport costs paid to airlines and other carriers

Payments to foreign owners of hotels and other facilities

The cost of goods and services imported for the tourist industry

Remittances sent home by foreign workers

Foreign debt relating to tourism

Payments to foreign companies to build tourist infrastructure

Leakages

Figure 13.9 Economic leakages

However, supporters of the development potential of tourism argue that:

- it benefits other sectors of the economy, providing jobs and income through the supply chain; this is called the **multiplier effect** (Figure 13.10) because jobs and money multiply as a result of tourism development

- it is an important factor in the balance of payments of many nations
- it provides governments with considerable tax revenues
- by providing employment in rural areas it can help to reduce rural–urban migration
- a major tourism development can act as a growth pole, stimulating the economy of the larger region
- it can create openings for small businesses in which start-up costs and barriers to entry are generally low
- it can support many jobs in the informal sector, where money goes directly to local people

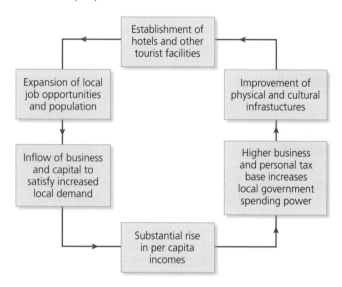

Figure 13.10 The multiplier effect of tourism

The environmental impact Revised

Tourism that does not destroy what it sets out to explore has come to be known as 'sustainable'. This is the kind of development that meets present needs without compromising the prospects of future generations. Following the 1992 Earth Summit in Rio de Janeiro, the WTTC and the Earth Council drew up an environmental checklist for tourist development, which included:

- waste minimisation
- re-use and recycling
- energy efficiency
- water management

In so many developing countries newly laid golf courses have taken land away from local communities while consuming large amounts of scarce freshwater. In both Belize and Costa Rica coral reefs have been blasted to allow for unhindered watersports. Education about the environment being visited is a key factor in reducing the environmental impact of tourism.

The environmental impact of tourism is not always negative. Landscaping and sensitive improvements to the built environment have significantly improved the overall quality of some areas. On a larger scale, tourist revenues can fund the designation and management of protected areas such as national parks and national forests.

The life cycle model of tourism Revised

Butler's model of the evolution of tourist areas (Figure 13.11) attempts to illustrate how tourism develops and changes over time:

- In the first stage the location is explored independently by a small number of visitors.
- If visitor impressions are good and local people perceive that real benefits are to be gained then the number of visitors will increase.
- In the development stage, holiday companies take control of organisation and management with package holidays becoming the norm.
- Eventually growth ceases as the location loses some of its former attraction. At this stage local people have become all too aware of the problems created by tourism.
- Finally, decline sets in, but efforts will be made to re-package the location which, if successful, may either stabilise the situation or result in renewed growth (rejuvenation).

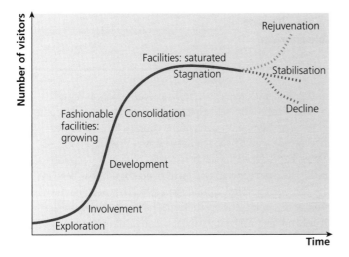

Figure 13.11 Butler's model of the evolution of tourism in a region

The model provides a useful summary of the stages that a number of holiday resorts, particularly in the Mediterranean, have been through. However, research has shown that it does not apply well to all locations. Prosser (1995) summarised the criticisms of the model:

- doubts on there being a single model of tourism development
- limitations on the capacity issue
- lack of empirical support for the concept
- limited practical use of the model

Also, it does not include the possible role of local and national governments in the destination country or the impact of, say, a low-cost airline choosing to add a destination to its network.

Recent developments in international tourism

Revised

More specialised types of tourism have become increasingly popular. One result of this has been the development of **niche tourism**. Niche market tour operators have increased in number to satisfy the rising demand for specialist holidays, which include:

- theme parks and holiday village enclaves
- gambling destinations
- cruising
- heritage and urban tourism
- wilderness and ecotourism
- religious tourism
- sports tourism

Niche tourism is tourism that deals in a specialised product.

Destination footprint is the environmental impact caused by an individual tourist on holiday in a particular destination.

Ecotourism

Environmental groups are keen to make travellers aware of their **destination footprint**. They are urging people to:

- 'fly less and stay longer'
- carbon-offset their flights
- consider 'slow travel'

Tourists are asked to consider the impact of their activities both on individual holidays but also in the longer term as well. For example, they may decide that every second holiday would be in their own country (not using air transport). It could also involve using locally run guesthouses and small hotels as opposed to hotels run by international chains. This enables more money to remain in local communities.

Virtually every aspect of the industry now recognises that tourism must become more sustainable. **Ecotourism** is at the leading edge of **sustainable tourism**.

> **Ecotourism** is a specialised form of tourism where people experience relatively untouched natural environments such as coral reefs, tropical forests and remote mountain areas, and ensure that their presence does no further damage to these environments.
>
> **Sustainable tourism** is tourism organised in such a way that its level can be sustained in the future without creating irreparable environmental, social and economic damage to the receiving area.

Case study — Ecotourism in Ecuador

International tourism is Ecuador's third largest source of foreign income after the export of oil and bananas. The number of visitors has increased substantially in recent years, both to the mainland and to the Galapagos Islands. The majority of tourists are drawn to Ecuador by the great diversity of flora and fauna. Much of the country is protected by national parks and nature reserves.

As visitor numbers began to rise Ecuador was anxious not to suffer the negative externalities of mass tourism, but to market 'quality' and 'exclusivity' instead, in as eco-friendly a way as possible. Ecotourism has helped to bring much-needed income to some of the poorest parts of the country. It has provided local people with a new, alternative way of making a living. As such it has reduced human pressure on ecologically sensitive areas. The main geographical focus of ecotourism has been in the Amazon rainforest around Tena, which has become the main access point. The ecotourism schemes in the region are usually run by small groups of indigenous Quichua Indians.

Now test yourself

Tested

9 Define tourism.

10 Why is it important to be aware of the carrying capacity of a tourist destination?

11 Describe Butler's life cycle model of tourism.

12 Why have more specialised types of tourism increased so much in popularity?

Answers on p.222

13.4 The management of a tourist destination: Jamaica

Jamaica's tourist industry

Revised

Jamaica is the third largest of the Caribbean islands. Tourism (Figure 13.12) originated in the latter part of the nineteenth century when a limited number of affluent people arrived to avoid the cold winters in the UK and North America. The first tourist hotels were built in Montego Bay and Port Antonio.

Advances in transportation made Jamaica a potential destination for an increasing number of people. In 2005 a total of 2,614,506 visited Jamaica. This comprised: 1,386,996 foreign nationals, 91,667 non-resident Jamaicans and 1,135,843 cruise passengers. The high or 'winter' season runs from mid-December to mid-April. The rainy season extends from May to November. About 25% of hotel workers are laid off during the off-season.

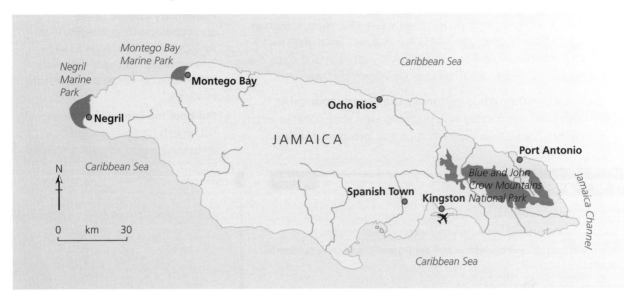

Figure 13.12 Tourist destinations in Jamaica

Jamaica is an example of a tourist area where there has been clear evidence of **growth** and **development**. As the industry has expanded, its linkages with other industries have developed as well. Tourism has become an increasingly vital part of Jamaica's economy in recent decades (Table 13.2). In 2007 the direct contribution to GDP was almost $1.2 billion; with all the indirect benefits added the figure was $3.8 billion. The contribution of tourism to total employment has also risen substantially. During the 1970s the Jamaican government introduced 'Jamaicanisation' policies designed to attract much-needed foreign investment in tourism. Policies included comparatively high wages and special industry taxes that went directly into social development, health care and education. Jamaica has been determined to learn from the 'mistakes' of other countries and ensure that the population will gain real benefits from the growth of tourism.

> **Growth** (of tourism) refers to the increase in numbers of tourists.
>
> **Development** (of tourism) refers to the expansion of tourism activities such as adventure tourism and ecotourism.

Table 13.2 The importance of the travel and tourist industry to the Jamaican economy, 2007

	Travel and tourism industry	Travel and tourism economy
GDP (% of total)	9.6	31.1
Employment	92,000	289,000

Jamaica's north coast, with its pleasant weather and white-sand beaches, is the centre of the island's tourist industry. The main resorts are Montego Bay, Ocho Rios and Port Antonio, although many tourists also visit the capital city, Kingston. While sun and sand are the main attractions, the island also has other attributes including dolphin parks, nature reserves, museums and galleries. There are excellent facilities for a range of sports. Jamaica's cuisine is an attraction for many visitors. There are many festivals and entertainment events during the year, often featuring Jamaica's native music, reggae.

Figure 13.12 shows the location of Jamaica's three national parks. A further six sites have been identified for future protection. The Jamaican government sees the designation of the parks as a positive environmental impact of tourism. Entry fees to the national parks pay for conservation. The two marine parks are attempting to conserve the coral reef environments off the west coast of Jamaica. They are at risk of damage from overfishing, industrial pollution and mass tourism.

Ecotourism and community tourism — Revised

Ecotourism is a developing sector of the industry with, for example, raft trips on the Rio Grande river increasing in popularity. Tourists are taken downstream in very small groups. The rafts, which rely solely on manpower, leave singly with a significant time gap between them to minimise any disturbance to the peace of the forest. Ecotourism is seen as the most sustainable form of tourist activity.

Considerable efforts are being made to promote **community tourism** so that more money filters down to the local population and small communities. Community tourism is seen as an important aspect of '**pro-poor tourism**'.

> **Community tourism** fosters opportunities at the community level for local people. Part of the tourist income gained is set aside for projects that provide benefits to the community as a whole.
>
> **Pro-poor tourism** results in increased net benefits for poor people.

Now test yourself — Tested

13 Describe the location of Jamaica.

14 Discuss the importance of tourism to the economy of Jamaica.

15 Briefly discuss the development of ecotourism and community tourism in Jamaica.

Answers on p.222

Exam-style questions

1 **(a)** Describe and explain the different types of international aid. [10]

(b) Critically discuss the impact of aid on receiving countries. [15]

2 **(a)** Examine the reasons for the growth of international tourism. [10]

(b) With reference to one tourist area, discuss the issues associated with the growth and development of tourism. [15]

Exam ready

14 Economic transition

14.1 National development

Employment structure and its role in economic development

People do hundreds of different jobs, which can be placed into four broad employment sectors:

- The **primary sector** produces raw materials from the land and the sea.
- The **secondary sector** manufactures primary materials into finished products.
- The **tertiary sector** provides services to businesses and to people.
- The **quaternary sector** uses high technology to provide information and expertise. Research and development is an important part of this sector.

The **product chain** can be used to illustrate the four sectors of employment. The food industry provides a good example (Figure 14.1). As an economy advances the proportion of people employed in each sector changes (Figure 14.2).

> The **product chain** is the full sequence of activities needed to turn raw materials into a finished product.

> **Expert tip**
>
> Secondary products are classed either as consumer goods (produced for sale to the public) or capital goods (produced for sale to other industries).

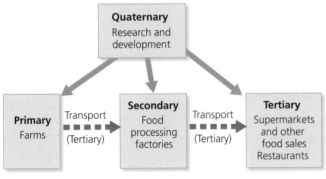

Figure 14.1 The food industry's product chain

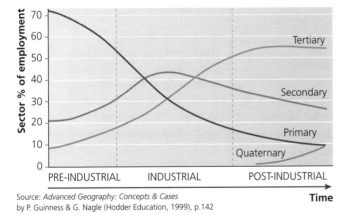

Source: *Advanced Geography: Concepts & Cases* by P. Guinness & G. Nagle (Hodder Education, 1999), p.142

Figure 14.2 The sector model

How employment structures vary

Developing countries

Many of the world's poorest countries are primary product dependent, making them very vulnerable to changes in world markets. Most of these people work in agriculture and many are **subsistence farmers**. In some regions of developing countries, mining, quarrying, forestry or fishing may dominate the economy. Work in mining in developing countries is often better paid than jobs elsewhere in the primary sector, but the working conditions are often very harsh.

In poor countries, higher-paid jobs in the secondary, tertiary and quaternary sectors are usually very few in number. The tertiary jobs that are available are often in the public sector. Public-sector jobs such as teaching, nursing and refuse collection are paid by the government. However, wages in these jobs are usually low as the funds available to the governments of developing countries are very limited.

> **Subsistence farming** is the most basic form of agriculture, where the produce is consumed entirely or mainly by the family who work the land or tend the livestock. If a small surplus is produced it may be sold or traded.

NICs

In newly industrialised countries employment in manufacturing has increased rapidly in recent decades. NICs attract foreign direct investment from transnational corporations in both the manufacturing and service sectors. The business environment in NICs is such that they also develop their own domestic companies. Such companies usually start in a small way, but some go on to reach a considerable size. The increasing wealth of NICs allows for greater investment in agriculture. This includes mechanisation, which results in a falling demand for labour on the land. So, as employment in the secondary and tertiary sectors rises, employment in the primary sector falls. Eventually, NICs may become so advanced that the quaternary sector begins to develop.

Developed countries

Developed countries are often referred to as **post-industrial societies**. Jobs in manufacturing industries have fallen for two reasons:

● Many manufacturing industries have moved to take advantage of lower costs in NICs and developing countries.
● Investment in robotics and other advanced technology has replaced much human labour in many manufacturing industries.

> A **post-industrialised society** is a developed country where far fewer people are now employed in manufacturing industries than in the past. Most people work in the tertiary sector, with an increasing number in the quaternary sector.

Global inequalities in social and economic well-being

Revised

Development and its traditional income measures

Development is a wide-ranging concept. It includes wealth, but also other important aspects of our lives such as health and freedom of speech. Development occurs when there are improvements to individual factors making up the quality of life. For example, development occurs in a low-income country when:

● local food supply improves due to investment in machinery and fertilisers
● levels of literacy improve throughout the country

The traditional indicator of a country's wealth has been **gross domestic product** (GDP). GDP is the total value of goods and services produced by a country in a year. A more recent measure is GNI (gross national income). To take account of the different populations of countries the **gross national income per capita** is often used. However, 'raw' or 'nominal' GNI data do not take into account the way in which the cost of living can vary between countries. For example, a dollar buys much more in China than it does is the USA. To account for this the **GNI at purchasing power parity** is calculated. The lowest GNI figures are concentrated in Africa and parts of Asia. The highest figures are in North America, the EU, Japan, Australia and New Zealand.

> **Gross national income (GNI)** comprises the total value of goods and services produced within a country (i.e. its gross domestic product), together with its income received from other countries (notably interest and dividends), less similar payments made to other countries.
>
> **GNI at purchasing power parity (PPP)** involves converting the GNI of a country into US dollars on the basis of how the value of the currency compares with that of other countries.

The development gap between the world's wealthiest and poorest countries is huge. However, a major limitation of GNI and other national data is that these are 'average' figures for a country, which tell us nothing about:

● the way in which wealth is distributed within a country
● how government invests the money at its disposal

The human development index

The **human development index (HDI)** devised by the United Nations (UN) in 1990 contains three variables:

- life expectancy
- educational attainment (adult literacy and combined primary, secondary and tertiary enrolment)
- GDP per capita (PPP$)

The actual figures for each of these three measures are converted into an index (Figure 14.3), which has a maximum value of 1.0 in each case. The three index values are then combined and averaged to give an overall human development index value. This also has a maximum value of 1.0. Every year the UN publishes the *Human Development Report* which uses the HDI to rank all the countries of the world according to their level of development. Every measure of development has merits and limitations. No single measure can provide a complete picture of the differences in development between countries. This is why the UN combines three measures of different aspects of the quality of life to arrive at a figure of human development for each country.

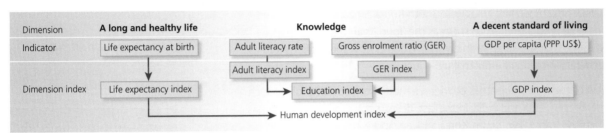

Figure 14.3 Constructing the HDI

Individual measures of development

There are many individual measures of socio-economic development, some of which have already been mentioned. Indicators not mentioned above include:

- number of people per doctor
- energy use per capita
- number of motor vehicles per 1000 people
- per capita food intake in calories
- televisions/refrigerators per 1000 population

The **infant mortality rate** is regarded as one of the most sensitive indicators of socio-economic progress. It is an important measure of health equity both between and within countries. The highest rates are clearly concentrated in Africa and southern Asia. Infant mortality generally compares well with other indicators of development.

Different stages of development

A reasonable division of the world in terms of stages of economic development is shown in Figure 14.4. The concept of **least developed countries** was first identified in 1968 by the UN. With 10.5% of the world's population these countries generate only one tenth of 1% of its income. Many of the least developed countries are in Sub-Saharan Africa. Others are concentrated in the poverty belt of Asia or are small island nations in the South Pacific. As the gap between the richest and poorest countries in the world widens, LDCs are being increasingly marginalised in the world economy. Least developed countries are usually dependent on one or a small number of exports for their survival.

> **Least developed countries (LDCs)** are the poorest and weakest economies in the developing world as identified by the UN.

> **Typical mistake**
>
> Students sometimes confuse LEDCs with LDCs. LDCs are the poorest subset of the LEDCs.

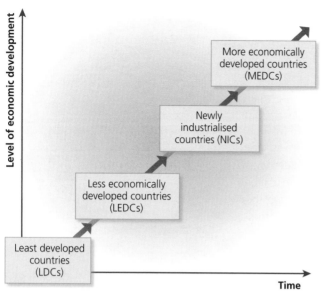

Figure 14.4 Stages of development

The first countries to become NICs were South Korea, Singapore, Taiwan and Hong Kong. The media referred to them as the 'four Asian tigers'. A 'tiger economy' is one that grows very rapidly. This group is now often referred to as the first generation of newly industrialised countries.

The success of these four countries provided a model for others to follow such as Malaysia, Brazil, China and India. In the last 15 years the growth of China has been particularly impressive. South Korea and Singapore have developed so much that many people now consider them to be developed countries.

> **Newly industrialised countries** are countries whose economic growth, particularly through manufacturing industry, has expanded significantly since the 1960s.

Explaining the development gap

There has been much debate about the causes of development. Detailed studies have shown that variations between countries are due to:

- physical geography:
  - Landlocked countries have generally developed more slowly than coastal ones.
  - Small island countries face considerable disadvantages in development.
  - Tropical countries have grown more slowly than those in temperate latitudes, reflecting the cost of poor health and unproductive farming in the former. However, richer, non-agricultural tropical countries such as Singapore do not suffer a geographical deficit of this kind.
  - A generous allocation of natural resources has spurred economic growth in a number of countries.
- economic policies:
  - Open economies that welcomed and encouraged foreign investment have developed faster than closed economies.
  - Fast-growing countries tend to have high rates of saving and low spending relative to GDP.
  - Institutional quality in terms of good government, law and order and lack of corruption generally result in a high rate of growth.
- demography:
  - Progress through demographic transition is a significant factor, with the highest rates of growth experienced by those nations where the birth rate has fallen the most.

Consequences of the development gap

The development gap has significant consequences for people in the most disadvantaged countries (Table 14.1). Development may not bring improvements in all four areas at first, but over time they should all show advances.

Table 14.1 Consequences of the development gap

Economic	Global integration is spatially selective: some countries benefit, others it seems do not. One in five of the world's population lives on less than a dollar a day, almost half on less than two dollars a day. Poor countries frequently lack the ability to pay for (1) food (2) agricultural innovation and (3) investment in rural development.
Social	More than 850 million people in poor countries cannot read or write. Nearly a billion people do not have access to clean water and 2.4 billion to basic sanitation. Eleven million children under 5 die from preventable diseases each year. The inability to combat the effects of HIV/AIDS is of huge concern.
Environmental	Poor countries have increased vulnerability to natural disasters. They lack the capacity to adapt to climate-change-induced droughts. Poor farming practices lead to environmental degradation. Often, raw materials are exploited with very limited economic benefit to poor countries and little concern for the environment.
Political	Poor countries that are low on the development scale often have non-democratic governments or they are democracies that function poorly. There is usually a reasonably strong link between development and improvement in the quality of government. In general, the poorer the country the worse the plight of minority groups.

Now test yourself

Tested

1 Why does the primary sector dominate employment in the poorest countries of the world?
2 Explain the changes in employment structure that have occurred in NICs.
3 The human development index includes education. Why is the level of education considered to be such an important measure of a country's development?
4 What were the reasons for the development of the first generation of newly industrialised countries?

Answers on pp.222–223

14.2 The globalisation of industrial activity

Global patterns of resources, production and markets

Revised

Globalisation is a recent phenomenon (post-1960), which is very different from anything the world has previously experienced. The modern global economy is more extensive and complicated than it has ever been before. There are many aspects of globalisation:

- economic
- urban
- social/cultural
- linguistic
- political
- demographic
- environmental

Globalisation is the increasing interconnectedness and interdependence of the world – economically, culturally and politically.

Global shift refers to the large-scale filter-down of economic activity from MEDCs to NICs and LEDCs.

Typical mistake

It is easy to think of globalisation as being only an economic phenomenon, but it has many other aspects, as identified in this section.

Transnational corporations and nation states are the two major decision-makers in the global economy. Nation states individually and collectively set the rules for the global economy but the bulk of investment is through TNCs, which are the main drivers of **global shift**. It is this process that has resulted in the emergence of an increasing number of NICs since the 1960s.

Economic globalisation

Figure 14.5 shows the main influences on the globalisation of economic activity. The factors responsible for economic globalisation are as follows:

● Until the post-1950 period the production process itself was mainly organised within national economies. This has changed rapidly in the last 50 years or so with the emergence of a **new international division of labour** (NIDL).

● The increasing complexity of international trade flows as this process has developed.

● Major advances in trade liberalisation under the World Trade Organization.

● The emergence of fundamentalist free-market governments in the USA and UK around 1980. The economic policies developed by these governments influenced policy-making in many other countries.

● The emergence of an increasing number of NICs.

● The integration of the old Soviet Union and its Eastern European communist satellites into the capitalist system. Now, no significant group of countries stands outside the free market global system.

● The opening up of other economies, particularly those of China and India.

● The deregulation of world financial markets.

● The 'transport and communications revolution', which has made possible the management of the complicated networks of production and trade that exist today.

> The **new international division of labour** divides production into different skills and tasks that are spread across regions and countries rather than within a single company.

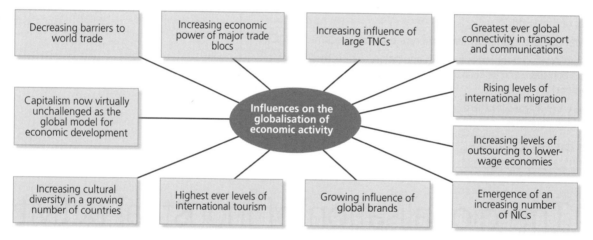

Figure 14.5 Globalisation trends

The advantages for economic activity in working at the global scale

Large companies in particular recognise many advantages in working at the global scale as opposed to the national or continental scale:

● Sourcing of raw materials and components on a global basis reduces costs.

● TNCs can seek out the lowest-cost locations for labour and other factors.

● High-volume production at low cost in countries such as China helps to reduce the rate of inflation in other countries and helps living standards to rise.

● Collaborative arrangements with international partners can increase the efficiency of operations considerably.

● Selling goods and services to a global market allows TNCs to achieve very significant economies of scale.

● Global marketing helps to establish brands with huge appeal all around the world.

Transnational corporations and foreign direct investment

Major TNCs and FDI flows

Investment involves expenditure on a project in the expectation of financial (or social) returns. **Transnational corporations** (TNCs) are the main source of **foreign direct investment** (FDI). As the rules regulating the movement of goods and investment have been relaxed in recent decades, TNCs have extended their global reach. FDI is not dominated by flows from core to periphery in the same way that it was even 20 years ago. Investment flows from NICs such as South Korea, Taiwan, China, India and Brazil have increased markedly.

TNCs have a substantial influence on the global economy in general and in the countries in which they choose to locate in particular. They play a major role in world trade in terms of what and where they buy and sell. A considerable proportion of world trade is intra-firm, taking place within TNCs. The organisation of the car giants exemplifies intra-firm trade with engines, gearboxes and other key components produced in one country and exported for assembly elsewhere. In 2009 the world's largest TNCs by revenue were led by Royal Dutch Shell, Exxon Mobil and Wal-Mart Stores. All three companies recorded revenue in excess of $400 billion.

> A **transnational corporation** is a firm that operates in more than one country.
>
> **Foreign direct investment** comprises overseas investments in physical capital by transnational corporations.

The development of TNCs

Figure 14.6 illustrates the sequential development of a transnational corporation, which begins with operation in the domestic market only. Large companies often reach the stage when they want to produce outside of their home country and take the decision to become transnational. The benefits of such a move include:

- cheaper labour, particularly in developing countries
- exploiting new resource locations
- circumventing trade barriers
- tapping market potential in other world regions
- avoidance of strict domestic environmental regulations
- exchange rate advantages

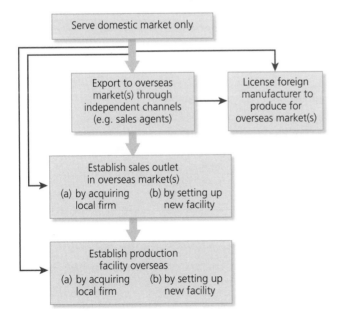

Figure 14.6 Sequential development of a transnational corporation

Contrasting spatial and organisational structures

TNCs vary widely in overall size and international scope. Variations include:

- the number of countries
- the number of subsidiaries
- the share of production accounted for by foreign activities

- the degree to which ownership and management are internationalised
- the division of research activities and routine tasks by country
- the balance of advantages and disadvantages to the countries in which they operate

Large TNCs often exhibit three organisational levels – headquarters, research and development and branch plants. Figure 14.7 shows the locational changes that tend to occur as TNCs develop over time.

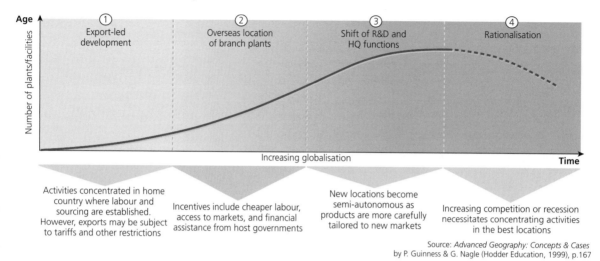

Source: *Advanced Geography: Concepts & Cases* by P. Guinness & G. Nagle (Hodder Education, 1999), p.167

Figure 14.7 The development of TNCs – locational changes

Case study **Nike**

Nike is the world's leading supplier of sports footwear, apparel and equipment and is one of the best-known global brands. It was founded in 1972. The company is an example of a vertical organisational structure across international boundaries, characterised by a high level of subcontracting activity. Nike does not make any shoes or clothes itself, but contracts out production to South Korean and Taiwanese companies.

Nike employs 650,000 contract workers in 700 factories worldwide. The company list includes 124 plants in China, 73 in Thailand, 35 in South Korea and 34 in Vietnam. More than 75% of the workforce are based in Asia. The majority of workers are women under the age of 25. Nike is a clear example of the new international division of labour (NIDL).

The emergence and growth of newly industrialised countries

In Asia four generations of NIC can be recognised in terms of the timing of industrial development and their current economic characteristics. Nowhere else in the world is the filter-down concept of industrial location better illustrated. When Japanese companies first decided to locate abroad in the quest for cheap labour, they looked to the most developed of their neighbouring counties, particularly South Korea and Taiwan. Most other countries in the region lacked the physical infrastructure and skill levels required by Japanese companies.

Companies from elsewhere in the developed world, especially the USA, also recognised the advantages of locating branch plants in such countries. As the economies of the first generation NICs developed, the level of wages increased, resulting in the following:

- Japanese and Western TNCs sought locations in second-generation NICs where improvements in infrastructure now satisfied their demands but where wages were still low.
- Indigenous companies from the first generation NICs moved routine tasks to their cheaper-labour neighbours such as Malaysia and Thailand.

With time, the process also included the third generation NICs, a significant factor in the recent very high growth rates in China and India. The least developed countries in the region, nearly all hindered by conflict of one sort or another at some time in recent decades, are now beginning to be drawn into the system. The recent high level of FDI into Vietnam makes it reasonable to think of the country as an example of a fourth-generation Asian NIC.

First-generation NICs

The reasons for the phenomenal rates of economic growth recorded in South Korea, Taiwan, Kong Hong and Singapore from the 1960s were:

- a good initial level of hard and soft infrastructure
- as in Japan previously, the land-poor NICs stressed people as their greatest resource, particularly through the expansion of education and training
- cultural traditions that revered education and achievement
- the Asian NICs becoming globally integrated at a 'moment of opportunity' in the structure of the world system
- all four countries having distinct advantages in terms of geographical location
- the ready availability of bank loans, often extended at government behest and at attractive interest rates

As their industrialisation processes have matured, the NICs have occupied a more intermediate position in the regional division of labour between Japan and other less-developed Asian countries.

Deindustrialisation

Revised

In the USA and Britain the proportion of workers employed in manufacturing has fallen from around 40% at the beginning of the twentieth century to less than half that now. Not a single developed country has bucked this trend, known as **deindustrialisation**, the causal factors of which are:

- technological change enabling manufacturing to become more capital intensive and more mobile
- the filter-down of manufacturing industry from developed countries to lower-wage economies
- the increasing importance of the service sector in the developed economies

> **Deindustrialisation** is the long-term absolute decline of employment in manufacturing.

Concerns about deindustrialisation have focused on the following:

- The traditional industries of the industrial revolution were highly concentrated, thus the impact of manufacturing decline has had severe implications in terms of unemployment and other social pathologies in a number of regions.
- The rapid pace of contraction of manufacturing has often made adjustment difficult.
- There are defence concerns if the production of some industries falls below a certain level.
- Some economists argue that over-reliance on services makes an economy unnecessarily vulnerable.
- Rather than being a smooth transition, manufacturing decline tends to concentrate during periods of economic recession.

It has been the revolution in transport and communications that that made such substantial filter-down of manufacturing to the developing world possible. Containerisation and the general increase in scale of shipping have cut the cost of the overseas distribution of goods substantially, while advances in telecommunications have made global management a reality.

In some cases whole industries have virtually migrated, as shipbuilding did from Europe to Asia in the 1970s. In others the most specialised work gets done in developed countries by skilled workers, and the simpler tasks elsewhere in the global supply chain.

14.3 Regional development

Income disparities within countries — Revised

The scale of income disparities *within* countries is often as much an issue as the considerable variations *between* countries. The Gini coefficient is a technique frequently used to show the extent of income inequality. It allows:

● analysis of changes in income inequality over time in individual countries
● comparison between countries

It is defined as a ratio with values between 0 and 1.0. A low value indicates a more equal income distribution while a high value shows more unequal income distribution. In 2007/8 the global gap ranged from 0.232 in Denmark to 0.707 in Namibia.

The Lorenz curve is a graphical technique that shows the degree of inequality that exists between two variables. It is often used to show the extent of income inequality in a population. A diagonal line represents perfect equality in income distribution. The further the curve is away from the diagonal line the greater the degree of income inequality.

In China, the income gap between urban residents and the huge farming population reached its widest level ever in 2008 as rural unemployment in particular rose steeply. The ratio between more affluent urban dwellers and their rural counterparts reached 3.36 to 1, up from 3.33 to 1 in 2007. This substantial income gap is a very sensitive issue in China as more and more rural people feel they have been left behind in China's economic boom. The size of the income gap is not just a political problem, but is also causing considerable national economic concern. Falling purchasing power in rural areas is hindering efforts to boost domestic consumer spending. The government wants to do this to help compensate for declining exports caused by the global recession.

Theory of regional disparities

The Swedish economist Gunnar Myrdal produced his **cumulative causation** theory in 1957. Figure 14.8 is a simplified version. Cumulative causation theory was set in the context of developing countries, but the theory can also be applied reasonably to more advanced nations as well. A three-stage sequence can be recognised:

● the pre-industrial stage, when regional differences are minimal
● a period of rapid economic growth characterised by increasing regional economic divergence (Figure 14.9)
● a stage of regional economic convergence when the significant wealth generated in the most affluent region(s) spreads to other parts of the country

Cumulative causation is the process whereby a significant increase in economic growth can lead to even more growth as more money circulates in the economy.

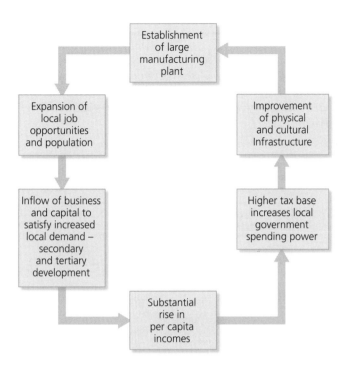

Figure 14.8 Simplified model of cumulative causation

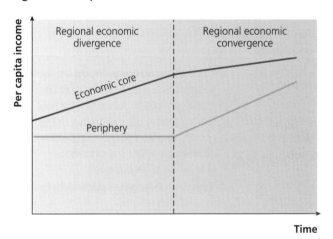

Figure 14.9 Regional economic divergence and convergence

In Myrdal's model, economic growth begins with the location of new manufacturing industry in a region. Once growth has been initiated in a dominant region spatial flows of labour, capital and raw materials develop to support it and the growth region undergoes further expansion by the cumulative causation process. A detrimental 'backwash effect' is transmitted to the less developed regions as skilled labour and locally generated capital are attracted away. Manufactured goods and services produced and operating under the scale economies of the economic 'heartland' flood the market of the relatively underdeveloped 'hinterland', undercutting smaller-scale enterprises in such areas.

However, increasing demand for raw materials from resource-rich parts of the hinterland may stimulate growth in other sectors of the economies of such regions. If the impact is strong enough to overcome local backwash effects a process of cumulative causation may begin, leading to the development of new centres of self-sustained economic growth. Such 'spread effects' are spatially selective.

The American economist Hirschman (1958) produced similar conclusions to Myrdal although he adopted a different terminology. Hirschman labelled the growth of the **economic core region** (heartland) as 'polarisation', which benefited from 'virtuous circles' or upward spirals of development, whereas the **periphery** (the hinterland) was impeded by 'vicious circles' or downward spirals. The term 'trickle-down' was used to describe the spread of growth from core to periphery.

The **economic core region** is the most highly developed region in a country, with advanced systems of infrastructure and high levels of investment resulting in high average income.

The **periphery** comprises the parts of a country outside the economic core region. The level of economic development in the periphery is significantly below that of the core.

Case study **Regional disparity in Brazil**

Southeast Brazil is the economic core region of the country. Over time the southeast has benefited from spatial flows of raw materials, capital and labour. The region grew rapidly through the process of cumulative causation. This process not only resulted in significant economic growth in the core, but also had a considerable negative impact on the periphery. The overall result was widening regional disparity. However, more recently, some parts of the periphery have benefited from spread effects (trickle-down) emanating from the core.

Such spread effects have been the result of a combination of market forces and regional economic policy. The south has been the most important recipient of spread effects from the southeast, but the other regions have also benefited to an extent. This process has caused the regional gap to narrow at times, but often not for very long. However, in Brazil income inequality still remains very wide.

Intra-urban variation: the growth of slums

Variations in the quality of life are extremely large in Brazil's big cities. The affluent apartment blocks in high-income areas contrast sharply with the slum conditions of the *favelas*. The numbers of people living in urban poverty are increased by a combination of economic problems, growing inequality and population growth, particularly growth due to in-migration (Figure 14.10).

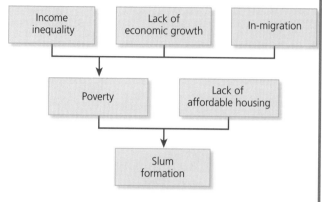

Figure 14.10 The formation of slums

Ethnicity and employment

The development gap often has an ethnic and/or religious dimension whereby some ethnic groups in a population have income levels significantly below the dominant group(s) in the same population. This is often the case with **indigenous populations**. This is invariably the result of discrimination, which limits the economic, social and political opportunities available to the disadvantaged groups. Examples include South Africa, Indonesia and Bolivia. Because of such obvious differences in status, tensions can arise between majority and minority groups, resulting in:

- social unrest
- migration
- new political movements

In South Africa the wide gap in income originated in the apartheid era, but since then it has proved extremely difficult to close for a variety of reasons. Political change often occurs well in advance of significant economic and social change. Inequality of wealth distribution is higher in Latin America than in any other part of the world. Indian and black people make up a third of the population, but have very limited parliamentary representation.

> The **indigenous population** are those people descended from the original ethnic group(s) to populate a country. Other ethnic groups migrating to that country at a later period of time may come to dominate the indigenous population in various ways.

Education

Education is a key factor in explaining disparities within countries. Those with higher levels of education invariably gain better paid employment. In developing countries there is a clear link between education levels and family size, with those with the least education having the largest families. Maintaining a large family usually means that saving is impossible and varying levels of debt likely. In contrast, people with higher educational attainment have smaller families and are thus able to save and invest more for the future. Such differences serve to widen rather than narrow disparities. Educational provision can vary significantly, not just by social class, but also by region.

Land ownership (tenure)

The distribution of land ownership has had a major impact on disparities in many countries. It can have a significant regional component. The greatest disparities tend to occur alongside the largest inequities in land ownership. The ownership of even a very small plot of land provides a certain level of security that those in the countryside without land cannot possibly aspire to.

In Brazil the distribution of land in terms of ownership has been a divisive issue since the colonial era. 44% of all arable land in Brazil is owned by just 1% of the nation's farmers, while 15 million peasants own little or no land. Many of these landless people are impoverished, roving migrants who have lost their jobs as agricultural labourers due to the spread of mechanisation in virtually all types of agriculture. At least a partial solution to the problem is land reform. This involves breaking up large estates and redistributing land to the rural landless. Although successive governments have vowed to tackle the problem, progress has been limited.

Now test yourself

8 Describe a technique that can be used to show the extent of income inequality.

9 Define the terms (a) economic core region, (b) periphery.

10 Explain how variations in educational attainment impact on regional development.

Answers on p.223

Tested

14.4 The management of development

Bolivia: managing the impact of globalisation — Revised

Bolivia (Figure 14.11) is South America's poorest country. It is one of only two South American countries that are landlocked.

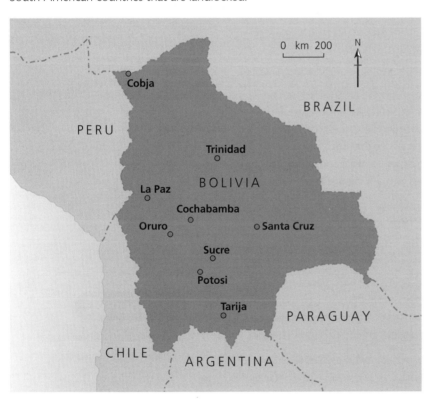

Figure 14.11 Bolivia

In the 1980s and 1990s the Bolivian government introduced free market reforms, which were required by the World Bank if Bolivia was to continue to receive aid. Although the Bolivian government had little choice it also wanted to link economic growth with equity so that poorer people would gain more benefit from Bolivia's participation in the global economy. The measures to achieve this included:

● a type of decentralisation called popular participation
● education reform to improve access to opportunities for the poor

However, very limited progress with these objectives led to frequent changes in government due to public disquiet. A significant change occurred in 2005 when Evo Morales of the Movement Toward Socialism (MAS) was elected as the country's first indigenous president. He was elected on a pledge to challenge the free market reforms that most people felt the country had been pressurised into adopting. There was widespread concern that these policies benefited large TNCs and the rich in Bolivia to the detriment of the poor and the environment.

Bolivia has a population of just under 10 million. The majority of the population are of indigenous origin. The indigenous peoples of Bolivia have had to endure a much lower quality of life than those of Spanish descent. The indigenous population has been particularly susceptible to:

- lack of economic opportunities in rural areas, where there are particularly high concentrations of indigenous peoples; this has resulted in large-scale migration to urban areas
- low employment rates in the formal sector and thus heavy reliance on the informal sector
- lack of access to land
- lack of access to basic social services (education, health, energy)
- continued discrimination and stereotyping

In May 2006 President Morales nationalised the country's gas and oil industry, a process known as **resource nationalisation**. The new government in Bolivia saw regaining control of the country's natural resources as a vital first step towards generating the revenue to achieve much needed development.

> **Resource nationalisation** is when a country decides to take part, or all, of certain natural resources under state ownership.

Bolivia is adopting a socialist model of regional commerce and cooperation as opposed to what it sees as 'US-backed free trade'. The government is trying to attract foreign investment while at the same time giving the state a larger role in managing the economy.

Bolivia has a problem with the USA's drug war in South America. The USA wants to end the production of coca and thus reduce cocaine production to zero. Although Bolivia is against the trade in illegal drugs it wants to preserve the legal market for coca-leaves and promote the export of legal coca products. The reduction in the coca crop has hit the incomes of many people on low incomes.

The actions of the USA in 'pressurising' Colombia and Peru into free-trade agreements has damaged Bolivian exports to these countries. 60% of Bolivia's main farm export, soya beans, goes to Colombia. Now Bolivia is concerned that cheap, subsidised US food will undercut much of Bolivia's market in Colombia.

In April 2006 Bolivia signed the people's trade agreement with Venezuela and Cuba. Cuba has agreed to take all of Bolivia's soya production as well as other farm products at market prices or better. Venezuela will also send oil to Bolivia to meet domestic shortages in production. Cuba has agreed to supply doctors.

Like many countries that rely heavily on the export of raw materials to earn foreign currency, Bolivia has a sizeable foreign debt. At the end of 2006, Bolivia owed $3.2 billion to foreign creditors. The Bolivian government is concerned by the number of people moving abroad to find work to earn money for their families. Limited employment in Bolivia is a major problem.

Now test yourself

11 What is resource nationalisation?

12 Why has Bolivia pursued this policy?

Answers on p.223

Tested ☐

Exam-style questions

1 (a) Discuss the characteristics of the primary, secondary, tertiary and quaternary sectors of an economy. [10]

   (b) With reference to examples, explain how the nature and importance of these sectors changes with economic development. [15]

2 (a) Describe and explain the role of transnational corporations in the global economy. [10]

   (b) Discuss the connections between industrial growth in some LEDCs/NICs and deindustrialisation in MEDCs. [15]]

Exam ready ☐

Now test yourself answers

Chapter 1

1 (a) Interception refers to precipitation that is collected and stored by vegetation. (b) Evaporation is the process by which water changes into a gas. (c) Infiltration refers to water that soaks into the ground.

2 Overland flow refers to water that was stored on the surface and then flows over the surface of the ground. In contrast, throughflow refers to water that has soaked into the ground and then moves through the soil to possibly reach a stream or river. In baseflow, the water seeps deep into the underlying bedrock and flows very slowly through the bedrock to eventually make its way to a stream or river. Generally, the lower down in the ground the flow, the slower it is.

3 Interception storage refers to the water that is trapped and caught by vegetation – trees intercept more precipitation than grasses and deciduous trees intercept more than coniferous ones, but only when they are in leaf.

4 Water that is stored on the ground may be evaporated (especially if it is warm and windy), or it may soak into the ground (infiltration), especially if the soil is quite porous, It may also flow over ground towards a stream (overland flow) especially when there is a steep gradient and impermeable rock.

5 On surfaces that have a high percentage vegetation cover infiltration rates are high. For example, on permanent pasture infiltration rates are over 50 mm/hour. In contrast, where there is less vegetation, infiltration rates are much lower (7 mm/hour on clean tilled land and 6 mm/hour on bare, crusted land). The data suggest that grazing reduces infiltration. On moderately grazed land infiltration is 19 mm/hour whereas on heavily grazed land it is 13 mm/hour. This may be because the vegetation is shorter (more grazed) and there is more bare ground (more trampling).

6 The Shannon has a fairly simple regime with higher discharge (over 20 l/s/km^2) in winter and lower discharge (below 10 l/s/km^2) in summer. Spring is characterised by falling discharge whereas autumn is characterised by rising discharge. In contrast, the peak flow of the Gloma is found during May and June (approximately 40 l/s/km^2). Lowest flows (less than 10 l/s/km^2) are found in winter and early spring. Autumn is characterised by falling discharges.

7 The Shannon lies in a temperate zone and is affected by low-pressure systems, especially in winter. These low-pressure systems push further north in the summer. Thus it receives mainly winter rainfall, and so discharges are higher in winter. The Gloma is found in sub-Arctic conditions – much of its catchment will be frozen throughout the winter months. Much of the precipitation falls as snow and this does not contribute much to river discharge. However, when the snow melts in late spring, there is a massive increase in river discharge associated with spring-melt or snow-melt.

8 River regime is the annual variation in the flow of a river, whereas a storm hydrograph shows the variations in the flow of a river associated with an individual storm or groups of storms – normally over a period of between 1 and 7 days.

9 In the natural channels and basin surface (Channel 3), there is a low peak flow and long lag time. In contrast, in the completely sewered basin with a highly impermeable surface (Channel 1) there is a very high peak flow and a very short lag time. The channel that is completely sewered but with a natural surface (Channel 2) has a very short lag time and an intermediate peak flow.

10 Hydraulic action is the force of air and water on the sides of rivers and in cracks. Attrition is the wearing away of the load carried by a river. It creates smaller, rounder particles. Abrasion is the wearing away of the bed and bank by the load carried by a river. Abrasion increases as velocity increases.

11 The load is transported downstream in a number of ways:
   - The smallest particles (silts and clays) are carried in suspension as the suspended load.
   - Larger particles (sands, gravels, very small stones) are transported in a series of 'hops' as the saltated load.
   - Pebbles are shunted along the bed as the bed or tracted load.
   - In areas of calcareous rock, material is carried in solution as the dissolved load.

12 Deposit – below 0.1 m/s; transport – c. 0.1–0.2 m/s; entrainment – c. 0.2–0.4 m/s; erosion – over 0.4 m/s

13 The recurrence interval refers to the regularity of a flood of a given size, for example a 50-year flood is a flood that is expected to occur, on average, once every 50 years.

14 Absolute drought is a period of at least 15 consecutive days with less than 0.2 mm of rainfall whereas partial drought is a period of at least 29 consecutive days during which the average daily rainfall does not exceed 0.2 mm.

15 In the sub-tropics around 20–30° N and S

Chapter 2

1 The albedo is the proportion of energy that is reflected back to the atmosphere.

2 Albedo varies with colour – light materials are more reflective than dark materials, and this can affect how much heat something takes in.

3 Mist is cloud at ground level and occurs when visibility is between 1000 m and 5000 m whereas fog is cloud at ground level and occurs when visibility is below 1000 m.

4 Radiation fog occurs when the ground loses heat at night by long-wave radiation and chills the air to its dew point. Advection fog is formed when warm air flows over a cold surface and its temperature is chilled to dew point.

5 A temperature inversion occurs when there is an increase in temperature with height, rather than a decrease.

6 Temperature inversions occur when there are relatively calm (high pressure) conditions and little wind causing the air to mix. As the cold air at the surface is dense, it will tend to stay at the surface. During the longer nights of winter there is even more time for the air near the surface to cool.

7 Temperature inversions are important as they influence air quality (pollution). Under high-pressure conditions and limited air movement, a temperature inversion will act like a lid on pollutants, causing them to remain in the lower atmosphere next to the Earth's surface.

8 The amount of insolation received by the urban and rural area is the same.

9 Heat loss due to evaporation is much greater in the rural area (24 units by day, 1 by night) compared with the urban area (1 unit by day and 1 unit by night).

10 The urban area gives up more heat from the ground to the surface than the rural area (22 units as opposed to 11 units). This could be due to artificial sources of heat, such as those used in a metro/underground system, heating from underground car parks, heat from basements etc.

11 Incoming solar radiation

12 December–January (summer solstice)

13 None

14 Sub-tropical high pressure

15 Land

16 It causes high pressure.

17 It causes low pressure.

18 Wind reversal

19 (a) Northwards into India (the southeast monsoon); (b) southwards towards Australia (the northeast monsoon)

20 Specific heat capacity is the amount of heat needed to raise the temperature of a body by 1°C.

21 Insolation has more atmosphere to pass through at the poles, and it is also less concentrated at the poles.

22 In evaporation, water changes from a liquid to a gas, and heat is absorbed.

23 When condensation occurs latent heat locked in the water vapour is released, causing a rise in temperature.

24 Sublimation occurs when ice (solid) changes to water vapour (gas) without any melting (water).

25 5 km

26 Condensation level is the altitude at which relative humidity reaches 100% and therefore condensation occurs, forming the base of clouds.

27 Uplift can be caused by convectional heating, frontal activity or relief (orographic uplift).

28 Instability refers to rising air (low pressure) whereas stability refers to descending air (high pressure).

29 Radiation fog occurs when the ground loses heat at night by long-wave radiation and chills the air to its dew point. Advection fog is formed when warm air flows over a cold surface and its temperature is chilled to dew point.

30 Mist and fog forms during stable air conditions.

31 There could be significant decreases in water availability in many areas, including the Mediterranean and southern Africa. Small mountain glaciers will disappear; extensive damage to coral will occur; however, there may be rising yields in some high latitudes.

32 As a result of increased temperatures, some glaciers and ice caps will have begun to melt, raising sea levels. In addition, increased atmospheric energy will mean that that there is increased storm frequency and storm intensity, also adding to the risk of coastal flooding.

33 International agreements, such as the Kyoto Protocol, give countries legally binding targets for cuts in emissions. Essentially, we need to release less greenhouse gas (burn less fossil fuel), develop alternative sources of energy, reduce deforestation, and increase energy conservation.

34 Urban climates are typically warmer, less windy, have more cloud cover and more rain.

35 The urban heat island is a pattern of temperature in large urban areas in which the maximum is found near the city centre, there is a plateau across the suburbs and a temperature cliff between the suburban and rural areas where temperatures are lower than the urban area.

36 Higher levels of air pollution make fog and smog more common in urban areas.

37 Urban microclimates are best observed during calm, high-pressure conditions because there is less mixing of air with neighbouring rural areas. During low-pressure conditions, winds mix the rural and urban air to give a more uniform composition.

Chapter 3

1 Oceanic crust is thinner and denser than continental crust; it is younger and is generally more basaltic. Continental crust is likely to contain more aluminium and is more granitic.

2 North American, Eurasian, South American, African, Pacific, Indo-Australian

3 (a) When lava cools on the sea floor, magnetic grains in the rock acquire the direction of the Earth's prevailing magnetic field at the time of cooling. (b) Sea-floor spreading refers to the expansion of the sea floor as a result of magma being forced up at ocean ridges, forcing the ocean plates apart.

4 At a mid-ocean ridge new oceanic crust is being formed and the ocean plate is being split apart. At a subduction zone dense oceanic crust is being forced under less dense continental crust.

5 Constructive and destructive plate margins are associated with volcanic activity.

6 Island arcs are a series of volcanic islands, formed in an arc-shape.

7 Island arc systems are formed when oceanic lithosphere is subducted beneath oceanic lithosphere.

8 Physical weathering is the breakdown of rocks in situ to produce smaller, angular fragments of the same rock, such as scree.

9 Freeze–thaw is most effective on jointed rocks in environments where moisture is plentiful and there are frequent fluctuations above and below freezing point.

10 In many desert areas daytime temperatures exceed 40°C whereas night-time ones are little above freezing. Rocks heat up by day and contract by night. As rock is a poor conductor of heat, stresses occur only in the outer layers. Only hot deserts have such extreme diurnal temperature ranges.

11 Physical weathering produces smaller, angular fragments of the same rock, such as scree. Chemical weathering forms new materials.

12 Humid tropical

13 Temperate and tropical wet–dry

14 Chemical weathering increases as temperature increases and as precipitation levels rise.

15 Mean annual temperature is of limited use. It is the number of cycles, for example above and below 0°C, that is important of freeze–thaw. Water is essential for freeze–thaw weathering.

16 Tors could be formed by deep chemical weathering under warm, wet conditions or by freeze–thaw processes in a periglacial climate, or a combination of the two.

17 Equifinality states that different processes could lead to the same end product.

18 Limestone is affected by carbonation–solution, deposition of calcite, freeze–thaw, fluvial erosion, glacial erosion and mass movements.

19 Swallow holes are formed by carbonation–solution, river erosion or by the collapse of a cave system.

20 Rock type largely determines resistance to denudation. However, regular jointing may increase the risk of denudation, by allowing more water to enter the rock.

21 A rockslide is a dry movement, whereas a mudflow is a much wetter movement.

22 Mass movements include any large-scale movement of the Earth's surface that is not accompanied by a moving agent such as a river, glacier or ocean wave.

23 Mass movements can be classified by speed, type of movement, water content and materials.

24 Shear strength refers to the internal resistance of a slope, whereas shear stress refers to the forces attempting to pull the material downslope.

25 Falls occur on steep slopes, especially on bare rock faces. Individual blocks become prised away. In a slide, the slope retains its shape, and a whole block slides downslope.

26 In a slide there is a constant velocity in the sliding mass, whereas in the slump, velocity is highest at the surface and lower at depth. The slide retains its shape; the slump forms a 'toe' that spreads out.

27 A rotational slide occurs when a mass of rock slides along a curved slip plane. An avalanche is a rapid mass movement of snow, ice, rock or earth, mainly in mountainous areas.

28 Avalanches can occur when newly fallen snow falls off older snow, especially in winter (a dry avalanche), while in spring partially melted snow can move (a wet avalanche), often triggered by skiing.

29 Rockfalls can be reduced through flattening slopes, slope drainage, reinforcement of rock walls by grouting with cement, anchor bolts and covering of walls with steel mesh.

30 Opencast mining can cause habitat destruction, dump failure/erosion, water pollution, air pollution, noise, vibration and visual intrusion.

31 Underground mining can cause dump failure/erosion, subsidence, water pollution, air pollution, visual intrusion and dereliction.

32 Natural causes include volcanic eruptions and acidification caused by coniferous plantations; human causes include burning coal and the use of fuel in vehicles.

33 There are many ways of reducing the impacts of acid rain. These include spreading lime dust on lakes and acidified soils, burning less fossil fuel, burning less sulfur-rich fossil fuel, capturing sulfur in power stations (filters, scrubbers).

Chapter 4

1 (a) The number of live births per 1000 population in a given year; (b) the number of deaths per 1000 population in a given year.

2 This is the difference between the crude birth rate and the crude death rate.

3 Unlike the crude birth rate the fertility rate takes account of both age structure and gender.

4 Demographic, social/cultural, economic, political

5 The number of deaths of infants under 1 year of age per 1000 live births in a given year.

6 In the developed world heart disease and cancer are the main causes of death. In the developing world, infectious and parasitic diseases account for over 40% of all deaths.

7 (a) The composition of a population, the most important elements of which are age and sex; (b) a bar chart, arranged vertically, that shows the distribution of a population by age and sex.

8 For every 100 people in the economically active population there are 80 people dependent on them.

9 The dependency ratio in developed countries is usually between 50 and 75, with the elderly forming an increasingly high proportion of dependents. Developing countries typically have higher dependency ratios, which can reach over 100, with young people making up the majority of dependents.

10 The historical shift of birth and death rates from high to low levels in a population.

11 Better nutrition; improved public health particularly in terms of clean water supply and efficient sewerage systems; medical advances

12 Natural decrease, with the birth rate lower than the death rate.

13 Two from: seen as too Eurocentric as it was based on the experience of western Europe; many developing countries may not follow the sequence set out in the model; it fails to take into account changes due to migration.

14 From 46 to 68 years

15 Very high life expectancy and very low fertility

16 Life expectancy, education and income.

17 The number of children who die before their 5th birthday per 1000 live births.

18 From almost 13 million to below 10 million a year.

19 The largest population that the resources of a given environment can support.

20 A sustainability indicator that takes into account the use of natural resources by a country's population.

21 Built-up land; fishing ground; forest; grazing land; cropland; carbon footprint

22 The capacity of an area or ecosystem to generate an ongoing supply of resources and to absorb its wastes.

23 In 1961 the global ecological footprint was well below world biocapacity. Since then the former has risen consistently and by the late 1980s it had equalled world biocapacity. In recent decades the global ecological footprint has increasingly exceeded biocapacity.

24 Four from: soil exhaustion, drought, floods, tropical cyclones, pests, disease.

25 Two from each column in Table 4.4

26 The population that achieves a given aim in the most satisfactory way.

27 Arithmetical progression: 1 – 2 – 3 – 4 – 5 – 6; geometrical progression: 1 – 2 – 4 – 8 –16 – 32

28 Famine, disease, war

29 When a government has a stated aim on an aspect of its population and it undertakes measures to achieve that aim.

30 Two from: the socioeconomic implications of population ageing; the decrease in the supply of labour; the long-term prospect of population decline.

31 1979

32 From 43.8/1000 to 13.6/1000

33 Three from: demographic ageing; an unbalanced sex ratio; a generation of 'spoiled' only children; a social divide as an increasing number of wealthy couples 'buy their way round' the legislation.

Chapter 5

1 The movement of people across a specified boundary, national or international, to establish a new permanent place of residence.

2 Migrations are embarked upon from an area of origin and are completed in an area of destination.

3 Shifting cultivation and nomadic pastoralism

4 There is an element of choice in impelled migration; there is no choice in forced migration.

5 A systems approach

6 Michael Todaro

7 'Closing up' at the point of origin; the actual cost of movement itself; the costs of 'opening up' at the point of destination

8 Any three factors from Table 5.1 (left column)

9 Any three factors from Table 5.1 (right column)

10 Censuses, population registers and social surveys

11 Employment, family reunion and marriage

12 Macro-level, meso-level and micro-level

13 (a) A region of concentrated economic development with advanced systems of infrastructure, resulting in high average income and relatively low unemployment. (b) A region of low or declining economic development characterised by low incomes, high unemployment, selective out-migration and poor infrastructure.

14 The specific circumstances of individual families and communities in terms of urban contact.

15 (a) Mainly young adults and their children; (b) to support education and the search for employment.

16 Money sent home to families by migrants working elsewhere

17 A deficit of young adults (and their children) in rural areas with significant out-migration. The reverse in receiving urban areas.

18 Nigeria

19 The period explanation, the regional restructuring explanation, and the de-concentration explanation

20 175 million – 1 in every 35 people

21 An internally displaced person is someone who has been forced to leave his/her home for reasons similar to a refugee but who remains in the same country.

22 Worsening tropical storms, desert droughts and rising sea levels

23 The extent of economic opportunities; the presence of family members or others of the same ethnic origin; the point of entry into the country

24 $397 billion

25 The dispersal of a people from their original homeland

26 As remittances from Nepalese migrants have increased, the poverty level in Nepal has decreased.

27 Any three from the five bullet points given on page 81.

28 Average income; unemployment rates; the growth of the labour force; the overall quality of life

29 The early 1980s

30 California and Texas

Chapter 6

1 People living in the countryside in farms, isolated houses, hamlets and villages. Under some definitions small market towns are classed as rural.

2 Less than 2%

3 The process of population decentralisation as people move from large urban areas to small urban settlements and rural areas.

4 More people in rural communities owning cars and shopping in urban areas; the lower prices and wider choice of large urban outlets.

5 The lack of affordable housing

6 (a) South Dorset; (b) largely due to competition from groups of different people including retirees and second-home owners.

7 60%

8 The increasing concentration of poverty in urban areas in developing countries due at least partly to high levels of rural–urban migration.

9 The former is the process whereby an increasing proportion of the population in a geographical area lives in urban settlements. The latter is the absolute increase in physical size and total population of urban areas.

10 Suburbanisation, counterurbanisation and reurbanisation

11 A mapping exercise by local government, which decides how land should be used in the various parts of a town or city.

12 The former involves complete clearance of existing buildings and site infrastructure, and constructing new buildings. The latter keeps the best elements of the existing urban environment and adapts them to new usages.

13 The Lower Lea Valley

14 1963 by Ruth Glass

15 The relative ease with which a place can be reached from other locations.

16 A city that is judged to be an important nodal point in the global economic system.

17 London and New York

18 Burgess

19 The zone in transition

20 Maximum accessibility

21 On the edge of the urban area

22 Urban density gradients

23 The long-term absolute decline of employment in manufacturing.

24 Two from the five bullet points on page 96

25 Core and frame

26 The complex pattern of different residential areas within a city reflecting variations in socio-economic status, which are mainly attributable to income.

27 18 million and 8110/km²

28 Heliopolis

29 Projeto Cingapura

30 16 million

31 The Greater Cairo Waste Water project

32 10

33 1911

34 38%

35 Social exclusion

36 The Hukou system

37 Balanced development

38 When rural settlements transform themselves into urban or quasi-urban entities with very little movement of population.

Chapter 7

1 A large body of air where the horizontal gradients (variation) of the main physical properties, such as temperature and humidity, are fairly gentle.

2 mT – tropical air that is hot and moist. cT – tropical continental air that is hot but dry.

3 The subtropical high or warm anticyclone is caused by cold air at the tropopause descending. This is the descending air of the Hadley cell.

4 The position of the high pressure alters in response to the position of the overhead sun and the seasonal drift of the ITCZ. It shifts about 5°N in July and about 5°S in January.

5 The most simple explanation for the monsoon is that it is a giant land–sea breeze. In summer, central Asia becomes very hot and a centre of low pressure develops. The air over the Indian Ocean and Australia is colder, and therefore denser, and sets up an area of high pressure. As air moves from high pressure to low pressure, air is drawn into Asia from over the oceans. This moist air is responsible for the large amount of rainfall that occurs in the summer months. By contrast, in the winter months the sun is overhead in the southern hemisphere. Australia is heated (forming an area of low pressure), whereas the intense cold over central Asia and Tibet causes high pressure. Thus in winter air flows outwards from Asia, bringing moist conditions to Australia.

6 Banjul has a very seasonal climate. It has a very wet season largely concentrated in August and September. Total rainfall is just over 1400 mm with a dry season from December to April. Average temperatures are high and range from around 22°C in December–January to about 26°C in June. The range of temperatures is greater in the dry season than the wet season (the effect of cloud cover).

7 Succession refers to the spatial and temporal changes in plant communities as they move towards a seral climax.

8 A plagioclimax refers to a plant community permanently influenced by human activity.

9 Owing to the year-round growing season, constantly high temperatures (above 26°C), availability of moisture (over 2000 mm/year) and availability of light, productivity remains very high. In addition, due to the high biodiversity of the ecosystem there are species exploiting every ecological niche, which makes the ecosystem productive.

10 In the tropical rainforest, the input of nutrients from weathering and precipitation is high because of the sustained warm wet conditions. Most of the nutrients are held in the biomass due to the continual growing season. Breakdown of nutrients is rapid under the warm wet conditions, and there is a relatively small store in the soil. Where vegetation has been removed, the loss of nutrients is high due to high rates of leaching and overland flow.

11 As a result of human activity, the store of nutrients in the biomass is reduced. This may occur due to harvesting of crops or deforestation for other land uses. Once the trees are removed the store of nutrients in the soil is rapidly reduced. If the trees are cut down and burnt for shifting cultivation, there may be an initial increase in the amount of nutrients in the soil, but after a year or so, nutrient availability decreases significantly.

12 (a) Savanna vegetation is adapted to drought by having deep tap roots to reach the water table, partial or total loss of leaves and/or sunken stomata on the leaves to reduce moisture loss. (b) Savanna fauna is adapted to drought by being migratory.

13 Fire helps to maintain the savanna as a grass community. It mineralises the litter layer, kills off weeds, competitors and diseases and prevents any trees from colonising relatively wet areas.

14 There may be deep weathering in many humid tropical areas because the availability of water and the consistently high temperatures maximise the efficiency of chemical reactions, and in the oldest part of the tropics these have

been operating for a very long period – often millions of years.

15 Tors are said to be joint-controlled as the original positioning and pattern of joints determines how weathered the granite will become. Where there are many joints the granite is more weathered; where there are few joints, the rock is less weathered and corestones may be preserved.

16 The two theories of bornhardt formation are: the stripping or exhumation theory – increased removal of regolith occurs so that unweathered rocks beneath the surface are revealed; parallel retreat – the retreat of valley sides occurs until only remnant inselbergs are left.

17 Cockpit karst is a landscape pitted with smooth-sided, soil-covered depressions and cone-like hills. Tower karst is a landscape characterised by upstanding rounded blocks set in a region of low relief.

18 The Popoluca's farming is diverse and mimics the natural rainforest. It is relatively stable and supports more people than plantations or ranching. It is a form of extensive farming.

19 This suggests that the vegetation is diverse but the soil is very infertile. Remove the vegetation and the rainforest becomes very infertile very quickly.

Chapter 8

1 Wavelength is the distance between two successive crests or troughs, and wave frequency is the number of waves per minute.

2 Swash is the movement of water up a beach, whereas backwash is the movement of water down a beach. Fetch is the distance of open water a wave travels over.

3 Plunging breakers are ones in which the shoreward face of the wave becomes almost vertical, curls over, and plunges forward and downward as an intact mass of water. In contrast, in surging breakers the front face and crest of the wave remain relatively smooth and the wave slides directly up the beach without breaking.

4 In wave refraction, the waves slow down and change shape, as they attempt to break parallel to the shore. Wave refraction will concentrate wave energy and therefore erosional activity on the headlands, while wave energy will be dispersed in the bays; hence deposition will tend to occur in the bays.

5 The coastal sediment system, or littoral cell system, is a simplified model that examines coastal processes and patterns in a given area. It operates at a variety of scales from a single bay, to a regional scale. Each littoral cell is a self-contained cell, in which inputs and outputs are balanced.

6 Bedload refers to sediment that is transported with continuous contact (traction or dragging) or discontinuous contact (saltation) with the seafloor. In traction, grains slide or roll along; this is a slow form of transport. In contrast, suspended load refers to sediment that is carried by turbulent flow and is generally held up by the water.

7 Swash-aligned coasts are orientated parallel to the crests of the prevailing waves. They are closed systems in terms of longshore drift transport, and the net littoral drift rates are zero. Drift-aligned coasts are orientated obliquely to the crest of the prevailing waves. The shoreline of drift-aligned coast is primarily controlled by longshore sediment transport processes. Drift-aligned coasts are open systems in terms of longshore drift transport. Therefore, spits, bars, and tombolos are all features of drift-aligned coasts.

8 In summer, owing to increased frequency of constructive waves, there is an increase in deposition and a build up of the beach, and a slightly convex profile. In contrast, in winter, there is an increase in destructive waves. Material is removed by the powerful backwash, and transferred offshore, leaving a concave profile.

9 Spits are bars of sand or shingle that are attached to the mainland at one end. The attached end is the proximal end and the unattached end is the distal end. The distal end may be recurved, as a result of wave refraction, cross currents and occasional storms. Behind the spit there may be a salt marsh.

10 A tombolo is a ridge that links an island to the mainland, whereas a bar is a ridge of material that is connected at both ends to the mainland.

11 Spits are formed by longshore drift along an irregular coastline. Longshore drift carries sediment downdrift. Here a change in the coastline occurs. Longshore drift is unable to respond immediately to the change and so continues to carry the sediment out and away from the shoreline. As a result of wave refraction, spits often become curved. Cross currents or occasional storm waves may assist this hooked formation.

12 Mud flats and salt marshes occur in low-energy environments behind spits, barrier islands and in estuaries. Rivers carrying clay into these areas may deposit their load, due to the low-energy conditions and the presence of sea water (clay flocculates in salt water).

13 Sand dunes form where there is a reliable supply of sand, strong onshore winds, a large tidal range and vegetation to trap the sand. Vegetation causes the wind velocity to drop, especially in the lowest few centimetres above the ground, and the reduction in velocity reduces energy and increases the deposition of sand. Vegetation is also required to stabilise dunes.

14 Temperature, depth of water, light, salinity, sediment, wave action and exposure to the air

15 Fringing reefs, barrier reefs and atolls

16 Any development that increases the standard of living of the population but does not compromise the needs of future generations.

17 Yachtsmen and fishermen

18 Land-use zoning has been used to manage conflicting land-uses.

Chapter 9

1 Primary hazards such as ground shaking and surface faulting, and secondary hazards such as ground failure and soil liquefaction, landslides and rockfalls, debris flows and mud flows, and tsunamis.

2 Measurement of small-scale ground-surface changes, such as uplift, subsidence and tilt, changes in rock stress,

micro-earthquake activity (clusters of small quakes), anomalies in the Earth's magnetic field, changes in radon gas concentration, and changes in electrical resistivity of rocks.

3 Tsunamis are generally caused by earthquakes (usually in subduction zones), but can be caused by volcanoes and landslides.

4 At present it is impossible to predict precisely where and when a tsunami will happen. However, once a tsunami has started an early warning system can be given to other areas.

5 Direct hazards (primary hazards) include pyroclastic flows, volcanic bombs (projectiles), lava flows, ash fallout, volcanic gases, nuee ardentes and earthquakes. Indirect hazards (secondary hazards) include atmospheric ash fall out, landslides, tsunamis, acid rainfall and lahars (mudflows).

6 It is possible to predict volcanic eruptions. Methods include seismometers to record swarms of tiny earthquakes that occur as the magma rises, chemical sensors to measure increased sulfur levels, lasers to detect the physical swelling of the volcano, and ultrasound to monitor low-frequency waves in the magma, resulting from the surge of gas and molten rock.

7 Experience – the more experience of environmental hazards, the greater the adjustment to the hazard. Material well-being – those who are better off have more choice. Personality – is the person a leader or a follower, a risk-taker or risk-minimiser. Education can also be a significant factor.

8 Options include: accept the hazard risk; try to modify the natural hazard process or the environment in which they live; leave the area altogether.

9 Increasing the slope angle, for instance cutting through high ground – slope instability increases with increased slope angle; placing extra weight on a slope, for instance new buildings – this adds to the stress on a slope; removing vegetation – roots bind the soil together and interception by leaves can reduce rainfall compaction; exposing rock joints and bedding planes can increase the speed of weathering.

10 High rainfall. Geologically the area is unstable – it has active volcanoes, such as Vesuvius, much high, steep land, and many fast-flowing rivers.

11 The clay soils of the mountains have been rendered dangerously loose by forest fires and deforestation. Houses have been built on hillsides identified as landslide zones. Many were built over a 2-metre-thick layer of lava formed by the eruption of Vesuvius in 79 AD.

12 Many avalanches occur between January and March because there has been a build up of snow and ice over winter and the temperatures are beginning to rise, causing melting.

13 Avalanches are mass movements of snow and ice.

14 Very heavy snowfalls can trigger avalanches. Newly fallen snow can fall off older snow, especially in winter, while in spring partially thawed snow moves, often triggered by skiing. Avalanches occur frequently on steep slopes over 22°, especially on north-facing slopes where the lack of sun inhibits the stabilisation of the snow. Loose avalanches, comprising of fresh snow, usually occur soon after a snowfall. By contrast, slab avalanches occur at a later date, when the snow has developed some cohesion. They are often started by a sudden rise in temperature, which causes melting.

15 The region experienced up to 2 m of snow in just 3 days.

16 Meteorologists have warned that global warming will lead to increased snow falls in the Alps, which are heavier and later in the season.

17 Managing hurricane impact is difficult. The unpredictability of tropical storm paths makes the effective management of tropical storms difficult. Moreover, the strongest storms do not always cause the greatest damage. People living in coastal areas face increased risks associated with tropical storms.

18 As yet there is no effective way of managing tornadoes. The best advice is to stay indoors and, if possible, underground. There is no proof that cloud seeding can or cannot change tornado potential in a thunderstorm.

19 The use of geo-materials has been very successful at Fraser's Hill, Pahang. As the site is fairly remote, higher transportation and labour cost would have contributed to the higher cost of construction of rock gabions. In contrast, the geo-materials that are abundantly available locally are relatively cheap to make or purchase. The geo-structures were non-polluting, required minimal post-installation maintenance, were visually attractive and could support greater biodiversity within the restored habitats. The geo-materials used in the project, such as straw wattles, biodegrade after about a year and become organic fertilisers for the newly established vegetation. After 18 months, the restored cut slopes were almost covered by vegetation, and there was no further incident of landslides. The geo-structures installed on site were cost-effective and visually attractive. The restored cut slopes were more stable and supported higher biological diversity. Overall, it was a major success.

Chapter 10

1 Rainfall effectiveness is the excess of rainfall (P) over potential evapotranspiration (E).

2 The deserts on the west coast of South America and southern Africa are caused by the presence of cold, upwelling currents. This limits the amount of moisture the air can hold.

3 The factors include belts of descending sub-tropical high pressure, continentality, rain-shadow effects and the presence of cold, upwelling ocean currents.

4 Rainfall variability (V) is expressed by:

$$V\,(\%) = \frac{\text{mean deviation from the average}}{\text{the average rainfall}} \times 100$$

5 There are two types of wind erosion. Deflation is the progressive removal of small material, leaving behind larger materials. This forms a stony desert or reg. In some cases, deflation may remove sand to form a deflation hollow. Abrasion is the erosion carried out by wind-borne particles. They act like sandpaper, smoothing surfaces, and exploiting weaker rocks. Most abrasion occurs within a metre of the surface, since this is where the largest, heaviest, most erosive particles are carried.

6 Barchan dunes are crescent-shaped and are found in areas where sand is limited but there is a constant wind supply. They have a gentle windward slope and a steep leeward slope up to 33°. In contrast, parabolic dunes have the opposite shape to barchans – they are crescent shaped, but point downwind. They occur in areas of limited vegetation or soil moisture.

7 Exotic or exogenous rivers are those that have their source in another wetter environment and then flow through a desert. Endoreic rivers are those that drain into an inland lake or sea. Ephemeral rivers are those that flow seasonally or after storms. Often they are characterised by high discharges and high sediment levels.

8 Alluvial fans form generally when a heavily sediment-laden river emerges from a canyon. The river, no longer confined to the narrow canyon, spreads out laterally, losing height, energy and velocity, so that deposition occurs. Larger particles are deposited first and finer materials are carried further away from the mountain. Pediments are the result of lateral planation, or could be the result of river deposition, similar to an alluvial fan.

9 Deserts have low rates of biomass productivity. This is due to the limited amount of organic matter caused by extremes of heat and lack of moisture.

10 To reduce water loss desert plants have many adaptations. A small surface area-to-volume ratio is an advantage. Water regulation by plants can be controlled by diurnal closure of stomata, and xerophytic plants have a mix of thick, waxy cuticles, sunken stomata and leaf hairs. The most drought-resistant plants are the succulents, including cacti, which possess well-developed storage tissues, small surface area-to-volume ratios, and rapid stomatal closure, especially during the daytime.

11 Desert soils, called aridisols, have a low organic content and are only affected by limited amounts of leaching. Soluble salts tend to accumulate in the soil either near the water table or around the depth of moisture percolation. As precipitation declines, this horizon occurs nearer to the surface. Desert soils also have a limited clay content.

12 Desertification is defined as land degradation in humid and semi-arid areas, i.e. non-desert (arid) areas. It involves the loss of biological and economic productivity and it occurs where climatic variability (especially rainfall) coincides with unsustainable human activities.

13 Natural causes of desertification include temporary periods of extreme drought and long-term climate change towards aridity.

14 Two human causes of desertification are over-grazing and over-cultivation. Over-grazing is the major cause of desertification worldwide. Vegetation is lost both in the grazing itself and in being trampled by large numbers of livestock. Overgrazed lands then become more vulnerable to erosion as compaction of the soils reduces infiltration, leading to greater runoff, while trampling increases wind erosion. Fencing, which confines animals to specific locations, and the provision of water points and wells have led to severe localised overgrazing. Boreholes and wells also lower the water table, causing soil salinisation. Over-cultivation leads to diminishing returns, where the yield decreases season by season, requiring an expansion of the areas to be cultivated simply to maintain the same return on the agricultural investment. Reducing fallow periods and introducing irrigation are also used to maintain output, but all these contribute to further soil degradation and erosion by lowering soil fertility and promoting salinisation.

15 The production of essential oils holds considerable potential as a form of sustainable agricultural development in the Eastern Cape. The raw materials are present and it is a labour-intensive industry that would utilise a large supply of unemployed and underemployed people. The essential oils industry has a number of advantages including a new or additional source of income for many people; it is local in nature; many plants are already known and used by the peoples as medicines, and are therefore culturally acceptable; in their natural state the plants are not very palatable or of great value and will not therefore be stolen; many species are looked upon as weeds – removing these regularly improves grazing potential as well as supplying raw materials for the essential oils industry.

Chapter 11

1 Very large farms; concentration on one (monoculture) or a small number of farm products; a high level of mechanisation; low labour input per unit of production; heavy usage of fertilisers, pesticides and herbicides; sophisticated ICT management systems; highly qualified managers; often owned by large agribusiness companies; often vertically integrated with food processing and retailing.

2 Intensive farming is characterised by high inputs per unit of land to achieve high yields per hectare. Extensive farming is where a relatively small amount of agricultural produce is obtained per hectare of land, so such farms tend to cover large areas of land. Inputs per unit of land are low.

3 'Wet' rice is grown in the fertile silt and flooded areas of the lowlands while 'dry' rice is cultivated on terraces on the hillsides.

4 Increasing the land under cultivation and increasing the yield per hectare.

5 The necessary high inputs of fertiliser and pesticide are costly in both economic and environmental terms; the problems of salinisation and waterlogged soils has increased with the expansion of irrigation; high chemical inputs have had a considerable negative effect on biodiversity; ill-health due to contaminated water and other forms of agricultural pollution; Green Revolution crops are often low in important minerals and vitamins.

6 Mainly sub-tropical or tropical maritime

7 The difficulties of making a living on marginal land; the removal of preferential treatment for bananas on the European market; crop disease.

8 The cultivation of non-traditional crops including sweet potatoes, yams and hot peppers for both domestic and international markets. Concentration on niche markets.

9 Mixed commercial farming

10 A range of traditional herbal remedies, spicy foods and refreshers, which have accounted for an increasing proportion of the farm's income in recent years.

11 Fixed transport costs include the equipment used to handle and store goods, and the costs of providing the transport system. Line haul costs are the costs of actually moving the goods and are largely composed of fuel costs and wages.

12 The wage rate is simply the hourly or weekly amount paid to employees, while the unit cost is a measure of productivity, relating wage rates to output.

13 The benefits that accrue to a firm by locating in an established industrial area. External economies of scale can be subdivided into urbanisation economies and localisation economies.

14 The clustering together and association of economic activities in close proximity to one another. Industrial agglomeration can result in companies enjoying the benefits of external economies of scale.

15 Industrial zones with special incentives set up to attract foreign investors, in which imported materials undergo some degree of processing before being re-exported.

16 17%, 28%, 52%

17 Liberalisation, deregulation and market orientation. Tariffs on imports were also significantly reduced along with other non-tariff trade barriers as a result of India's membership of the World Trade Organization (WTO).

18 Labour costs are considerably lower; a number of developed countries have significant ICT skills shortages; India has a large and able English-speaking workforce.

Chapter 12

1 The relative contribution of different energy sources to a country's energy production/consumption.

2 Coal gasification, clean coal technologies and the extraction of unconventional natural gas

3 Large dams and power plants can have a huge negative visual impact on the environment; the obstruction of the river affects aquatic life; there is deterioration in water quality; large areas of land may need to be flooded to form the reservoir behind the dam; submerging large forests without prior clearance can release significant quantities of methane, a greenhouse gas.

4 In developing countries about 2.5 billion people rely on fuelwood (including charcoal) and animal dung for cooking. Fuelwood accounts for just over half of global wood production. It provides much of the energy needs for Sub-Saharan Africa. It is also the most important use of wood in Asia.

5 2009

6 Energy conservation; placing a strong emphasis on domestic resources; diversified energy development; environmental protection; mutually beneficial international cooperation

7 Strategic petroleum reserve

8 The Three Gorges Dam has been a major part of China's policy in reducing its reliance on coal. When totally complete the generating capacity of the dam will be a massive 22,500 MW. The Dam supplies Shanghai and Chongqing in particular with electricity.

9 The deterioration of the environment through depletion of resources such as air, water and soil.

10 Two from: increasingly strict environmental legislation; industry has spent increasing amounts on research and development to reduce pollution; the relocation of the most polluting activities to the emerging market economies.

11 Water-stressed areas occur where water supply is below 1700 m^3 per person per year. Water-scarce areas occur where water supply falls below 1000 m^3 per person per year.

12 The main impacts of agro-industrialisation have been: deforestation; land degradation and desertification; salinisation and contamination of water supplies; air pollution; increasing concerns about the health of long-term farm workers; landscape change; declines in biodiversity.

13 Legally recognised common property resource management organisations in Namibia's communal lands.

14 1996

15 From less than 20,000 km^2 in 1998 to more than 100,000 km^2 in 2005

16 Contracts with tourism companies; selling hunting concessions; managing campsites; selling wildlife to game ranchers; selling crafts

Chapter 13

1 Visible trade involves items that have a physical existence and can actually be seen. Thus raw materials (primary products), such as oil and food, and manufactured goods (secondary products), such as cars and furniture, are items of visible trade. Invisible trade is trade in services, which include travel and tourism, and business and financial services.

2 A range of factors influence the volume, nature and direction of global trade, including: resource endowment, comparative advantage, locational advantage, investment, historical factors, the terms of trade, changes in the global market and trade agreements.

3 There is a strong relationship between trade and economic development. In general, countries that have a high level of trade are richer than those that have lower levels of trade. Countries that can produce goods and services in demand elsewhere in the world will benefit from strong inflows of foreign currency and from the employment its industries provide. Foreign currency allows a country to purchase abroad goods and services that it does not produce itself or produce in large enough quantities.

4 The fair trade system operates as follows: small-scale producers group together to form a cooperative with high social and environmental standards; these cooperatives deal directly with companies such as Tesco in MEDCs; MEDC companies pay significantly over the world market price for the products traded; the higher price achieved by the LEDC cooperatives provides both a better standard of living and some money to reinvest in their farms.

5 (a) Debt is that part of the total debt in a country owed to creditors outside the country. (b) The debt service ratio

is the ratio of debt service payments of a country to that country's export earnings.

6 The colonising powers often left their former colonies with high and unfair levels of debt when they became independent. Such debts were often at very high interest rates.

7 It began with the Arab-Israeli war of 1973–74, which resulted in a sharp increase in oil prices. Governments and individuals in the oil-producing countries invested the profits from oil sales in the banks of MEDCs. Eager to profit from such a high level of investment, these banks offered relatively low interest loans to poorer countries. They were encouraged to exploit raw materials and grow cash crops so that they could pay back their loans with profits made from exports. However, periods of recession in the 1980s and 1990s led to rising inflation and interest rates. At the same time crop surpluses resulted in a fall in prices. As a result the demand for exports from developing countries fell and export earnings declined significantly as a result. These factors, together with oil price increases, left many developing countries unable to pay the interest on their debts.

8 This is aid supplied by a donor country whereby the level of technology and the skills required to service it are properly suited to the conditions in the receiving country. This makes it much more likely that such aid will be sustainable and not reliant on further inputs from donor countries.

9 Tourism is travel away from the home environment for leisure, recreation and holidays; to visit friends and relatives; and for business and professional reasons.

10 The carrying capacity is the number of tourists a destination can take without placing too much pressure on local resources and infrastructure. If the carrying capacity of a tourist location is soon to be reached, important decisions have to be made. Will measures be taken to restrict the number of tourists, to remain within the carrying capacity, or are their possible management techniques that will allow the carrying capacity to be increased, but continue to be sustainable?

11 Butler's model of the evolution of tourist areas attempts to illustrate how tourism develops and changes over time. In the first stage the location is explored independently by a small number of visitors. If visitor impressions are good and local people perceive that real benefits are to be gained then the number of visitors will increase as the local community becomes actively involved in the promotion of tourism. In the development stage, holiday companies from the developed nations take control of organisation and management, with package holidays becoming the norm. Eventually growth ceases as the location loses some of its former attraction. At this stage local people have become all too aware of the problems created by tourism. Finally, decline sets in, but because of the perceived economic importance of the industry efforts will be made to re-package the location, which, if successful, may either stabilise the situation or result in renewed growth (rejuvenation).

12 An important factor seems to be a general re-assessment of the life–work balance. An increasing number of people are determined not to let work dominate their lives. One result of this has been the development of specialised or niche tourism such as cruising, gambling destinations and sports tourism. Rising real incomes and more leisure time have been important influential factors.

13 Jamaica is located in the Caribbean Sea approximately 90 miles south of Cuba.

14 Tourism's direct contribution to GDP in 2007 amounted to almost $1.2 billion. Adding all the indirect economic benefits increased the figure to almost $3.8 billion or 31.1% of total GDP. Direct employment in the industry amounted to 92,000 but the overall figure, which includes indirect employment, is over three times as large. In the most popular tourist areas the level of reliance on the industry is extremely high.

15 Ecotourism is a developing sector of the industry with, for example, raft trips on the Rio Grande river increasing in popularity. Tourists are taken downstream in very small groups. The rafts, which rely solely on manpower, leave singly with a significant time gap between them to minimise any disturbance to the peace of the forest. Ecotourism is seen as the most sustainable form of tourist activity. Considerable efforts are being made to promote community tourism so that more money filters down to the local population and small communities. The Sustainable Communities Foundation through Tourism (SCF) program has been particularly active in central and southwest Jamaica. Community tourism is seen as an important aspect of 'pro-poor tourism'.

Chapter 14

1 The poorest countries of the world have more than 70% of their employment in the primary sector. Lack of investment in general means that agriculture and other aspects of the primary sector are very labour intensive and jobs in the secondary and tertiary sectors are limited in number.

2 In NICs employment in manufacturing has increased rapidly in recent decades. NICs have reached the stage of development whereby they attract foreign direct investment from transnational corporations in both the manufacturing and service sectors. The business environment in NICs is such that they also develop their own domestic companies. Such companies usually start in a small way, but some go on to reach a considerable size. Both processes create employment in manufacturing and services. The increasing wealth of NICs allows for greater investment in agriculture. This includes mechanisation, which results in falling demand for labour on the land. So, as employment in the secondary and tertiary sectors rises, employment in the primary sector falls. Eventually, NICs may become so advanced that the quaternary sector begins to develop.

3 Quality education generally, and female literacy in particular, are central to development. The World Bank has concluded that improving female literacy is one of the most fundamental achievements for a developing nation to attain, because so many aspects of development depend on it. For example, there is a very strong relationship between the extent of female literacy and infant and child mortality rates. People who are literate are able to access medical and other information that will help them to a higher quality of life compared with those who are illiterate.

4 A good initial level of infrastructure; a skilled but relatively low-cost workforce; cultural traditions that revered education and achievement; governments welcomed foreign direct investment (FDI) from transnational corporations; all four countries had distinct advantages in terms of geographical location; the ready availability of bank loans, often extended at government behest and at attractive interest rates.

5 The new international division of labour (NIDL) is characterised by a significant change in the geographical pattern of specialisation, with the fragmentation of many production processes across national boundaries. This divides production into different skills and tasks that are spread across regions and countries rather than within a single company.

6 The main factors are: the emergence of a new international division of labour; the increasing complexity of international trade flows as this process has developed; major advances in trade liberalisation under the World Trade Organization; the emergence of fundamentalist free-market governments in the USA and UK around 1980, which influenced policy-making in many other countries; the emergence of an increasing number of NICs; the integration of the old Soviet Union and its Eastern European communist satellites into the capitalist system; today no significant group of countries standing outside the free market global system; the opening up of other economies, particularly those of China and India; the deregulation of world financial markets; the 'transport and communications revolution' that has made possible the management of the complicated networks of production and trade that exist today.

7 The main causal factors of deindustrialisation have been: technological change enabling manufacturing to become more capital intensive and more mobile; the filter-down of manufacturing industry from developed countries to lower-wage economies; the increasing importance of the service sector in the developed economies.

8 The Gini coefficient is a technique frequently used to show the extent of income inequality. It allows analysis of changes in income inequality over time in individual countries and comparison between countries. It is defined as a ratio with values between 0 and 1. A low value indicates a more equal income distribution while a high value shows more unequal income distribution.

9 (a) The most highly developed region in a country with advanced systems of infrastructure and high levels of investment resulting in high average income. (b) The parts of a country outside the economic core region. The level of economic development in the periphery is significantly below that of the core.

10 Education is a significant factor in explaining regional disparities. Those with higher levels of education invariably gain better paid employment. In developing countries there is a clear link between education levels and family size, with those with the least education having the largest families. Maintaining a large family usually means that saving is impossible and varying levels of debt likely. In contrast, people with higher educational attainment have smaller families and are thus able to save and invest more for the future. Such differences serve to widen, rather than narrow, disparities.

11 When a country decides to take part, or all, of one or a number of natural resources under state ownership.

12 To regain an important aspect of its sovereignty and to use the expected increase in national income to combat inequality and poverty. The Bolivian government believed that too much of the profits from the oil and gas industry were going to foreign TNCs with too little of the wealth filtering down to the average person in Bolivia.